Lecture Notes
in Business Information Processing   333

Stanisław Wrycza · Jacek Maślankowski (Eds.)

Information Systems: Research, Development, Applications, Education

11th SIGSAND/PLAIS EuroSymposium 2018
Gdansk, Poland, September 20, 2018
Proceedings

Springer

Editors
Stanisław Wrycza
Department of Business Informatics
University of Gdansk
Gdansk
Poland

Jacek Maślankowski
Department of Business Informatics
University of Gdansk
Gdansk
Poland

ISSN 1865-1348   ISSN 1865-1356 (electronic)
Lecture Notes in Business Information Processing
ISBN 978-3-030-00059-2   ISBN 978-3-030-00060-8 (eBook)
https://doi.org/10.1007/978-3-030-00060-8

Library of Congress Control Number: 2018952670

This Springer imprint is published by the registered company Springer Nature Switzerland AG
The registered company address is: Gewerbestrasse 11, 6330 Cham, Switzerland

Preface

EuroSymposium is already in its 11th year. This cyclical, annual event is changing and extending its thematic scope in accordance with the rapid progress being made within the field of information systems knowledge and applications, with its currently emerging areas, in the era of digital transformation. The new topics fall within the sphere of information systems and include Blockchain, Social Networks, Mobile Systems, the Internet of Things, Big Data, Machine Learning, Industry 4.0, User Experience, and several others. All of them have a strong influence on the EuroSymposium subject matter, changing it fundamentally as compared with the first EuroSymposia.

The objective of the PLAIS/SIGSAND EuroSymposium 2018 was to promote and develop high-quality research on all issues related to research, development, applications, and education in information systems (IS), often referred to in Europe as business informatics. It provided a forum for IS researchers and practitioners in Europe and beyond to interact, collaborate, and develop this field. The EuroSymposia were initiated by Prof. Keng Siau as the SIGSAND – Europe Initiative. Previous EuroSymposia were held at:

- University of Galway, Ireland – 2006
- University of Gdansk, Poland – 2007
- University of Marburg, Germany – 2008
- University of Gdansk, Poland – 2011
- University of Gdansk, Poland – 2012
- University of Gdansk, Poland – 2013
- University of Gdansk, Poland – 2014
- University of Gdansk, Poland – 2015
- University of Gdansk, Poland – 2016
- University of Gdansk, Poland – 2017

The accepted papers of the EuroSymposia held in Gdansk have been published in:

- 2nd EuroSymposium 2007: Bajaj, A., Wrycza, S. (eds.): Systems analysis and design for advanced modeling methods: best practises, information science reference, IGI Global, Hershey, New York (2009)
- 4th EuroSymposium 2011: Wrycza, S. (ed.): Research in Systems Analysis and Design: Models and Methods. LNBIP, vol. 93. Springer, Berlin (2011)
- Joint Working Conferences EMMSAD/EuroSymposium 2012 held at CAiSE'12: Bider, I., Halpin, T., Krogstie, J., Nurcan, S., Proper, E., Schmidt, R., Soffer, P., Wrycza, S. (eds.): Enterprise, Business-Process and Information Systems Modeling. LNBIP, vol. 113. Springer, Berlin (2012)
- 6th SIGSAND/PLAIS EuroSymposium 2013: Wrycza, S. (ed.): Information Systems: Development, Learning, Security. LNBIP, vol. 161. Springer, Berlin (2013)

- 7th SIGSAND/PLAIS EuroSymposium 2014: Wrycza, S. (ed.): Information Systems: Education, Applications, Research. LNBIP, vol. 193. Springer, Berlin (2014)
- 8th SIGSAND/PLAIS EuroSymposium 2015: Wrycza, S. (ed.): Information Systems: Development, Applications, Education. LNBIP, vol. 232, Springer, Berlin (2015)
- 9th SIGSAND/PLAIS EuroSymposium 2016: Wrycza, S. (ed.): Information Systems: Development, Research, Applications, Education. LNBIP, vol. 264, Springer, Berlin (2016)
- 10th SIGSAND/PLAIS EuroSymposium 2017: Wrycza, S., Maślankowski, J. (ed.): Information Systems: Development, Research, Applications, Education. LNBIP, vol. 300. Springer, Berlin (2017)

There were three organizers of the 11th EuroSymposium on Systems Analysis and Design:

- SIGSAND – Special Interest Group on Systems Analysis and Design of AIS
- PLAIS – Polish Chapter of AIS
- Department of Business Informatics of the University of Gdansk, Poland

SIGSAND is one of the most active SIGs with quite a substantial record of contributions for AIS. It provides services such as annual American and European Symposia on SIGSAND, research and teaching tracks at major IS conferences, listserv, and special issues in journals.

The Polish Chapter of the Association for Information Systems (PLAIS) was established in 2006 as the joint initiative of Prof. Claudia Loebbecke, former President of AIS, and Prof. Stanislaw Wrycza, University of Gdansk, Poland. PLAIS co-organizes international and domestic IS conferences and gained the title of Outstanding Chapter of the Association for Information Systems for the year 2017.

The Department of Business Informatics of the University of Gdansk conducts intensive teaching and research activities. Some of its academic manuals are bestsellers in Poland and the department is also is active internationally. The most significant conferences organized by the department were: the Xth European Conference on Information Systems, ECIS 2002, and the International Conference on Business Informatics Research, BIR 2008. The department is a partner of the ERCIS consortium – European Research Center for Information Systems. The students of Business Informatics of the University of Gdansk have been awarded several times at the annual AIS Student Chapters Competition.

EuroSymposium 2018 received 36 papers from 11 different countries. The submission and review process was supported by the Online Conference Service (OCS) hosted by Springer. The members of the International Program Committee carefully evaluated the submissions, selecting 14 papers for publication in this LNBIP volume. Therefore, EuroSymposium 2018 had an acceptance rate of 40%, with submissions divided into the following four groups:

- Systems Development and Engineering
- Systems Acceptance and Usability
- Internet of Things and Big Data
- Healthcare IT

The accepted papers reflect the current trends in information systems.

I would like to express my thanks to all authors, reviewers, advisory board members, and International Program Committee and Organizational Committee members for their support, effort, and time. They have made possible the successful accomplishment of EuroSymposium 2018.

July 2018   Stanisław Wrycza

Organization

General Chair

Stanislaw Wrycza   University of Gdansk, Poland

Organizers

- SIGSAND is the Association for Information Systems (AIS) Special Interest Group on Systems Analysis and Design
- The Polish Chapter of Association for Information Systems - PLAIS
- Department of Business Informatics at University of Gdansk

Advisory Board

Wil van der Aalst   Eindhoven University of Technology, The Netherlands
David Avison   ESSEC Business School, France
Joerg Becker   European Research Center for Information Systems, Germany
Jane Fedorowicz   Bentley University, USA
Alan Hevner   University of South Florida, USA
Helmut Krcmar   Technical University of Munich, Germany
Claudia Loebbecke   University of Cologne, Germany
Keng Siau   Missouri University of Science and Technology, USA
Roman Slowiński   Polish Academy of Sciences, Poland

International Program Committee

Özlem Albayrak   Bilkent University, Turkey
Eduard Babkin   Higher School of Economics, Moscow, Russia
Witold Chmielarz   University of Warsaw, Poland
Helen Dudycz   Wroclaw University of Economics, Poland
Tomasz Dzido   University of Gdansk, Poland
Marco de Marco   Università Cattolica del Sacro Cuore, Italy
Petr Doucek   University of Economics, Prague, Czech Republic
Zygmunt Drążek   University of Szczecin, Poland
Dariusz Dziuba   University of Warsaw, Poland
Peter Forbrig   University of Rostock, Germany
Bogdan Franczyk   University of Leipzig, Germany
Jerzy Gołuchowski   University of Economics in Katowice, Poland
Marta Indulska   The University of Queensland, Australia
Piotr Jedrzejowicz   Gdynia Maritime University, Poland

Dorota Jelonek	Czestochowa University of Technology, Poland
Bjoern Johansson	Lund University, Sweden
Kalinka Kaloyanova	Sofia University, Bulgaria
Karlheinz Kautz	Royal Melbourne Institute of Technology (RMIT) University, Australia
Marite Kirikova	Riga Technical University, Latvia
Jolanta Kowal	University of Wroclaw, Poland
Stanisław Kozielski	Silesian University of Technology, Poland
Henryk Krawczyk	Gdansk University of Technology, Poland
Kyootai Lee	Sogang University, South Korea
Tim A. Majchrzak	University of Agder, Norway
Marek Miłosz	Lublin University of Technology, Poland
Ngoc-Thanh Nguyen	Wroclaw University of Technology, Poland
Marian Niedźwiedziński	University of Lodz, Poland
Nikolaus Obwegeser	Aarhus University, Denmark
Cezary Orłowski	WSB University in Gdansk, Poland
Joanna Paliszkiewicz	Warsaw University of Life Sciences, Poland
Nava Pliskin	Ben-Gurion University of the Negev, Israel
Isabel Ramos	The University of Minho, Portugal
Vaclav Repa	University of Economics, Prague, Czech Republic
Michael Rosemann	Queensland University of Technology, Australia
Kurt Sandkuhl	University of Rostock, Germany
Thomas Schuster	Pforzheim University, Germany
Marcin Sikorski	Gdansk University of Technology, Poland
Janice C. Sipior	Villanova University, USA
Andrzej Sobczak	Warsaw School of Economics, Poland
Piotr Soja	Cracow University of Economics, Poland
Reima Suomi	University of Turku, Finland
Jakub Swacha	University of Szczecin, Poland
Pere Tumbas	University of Novi Sad, Serbia
Catalin Vrabie	National University, Romania
Yinglin Wang	Shanghai University of Finance and Economics, China
H. Roland Weistroffer	Virginia Commonwealth University, USA
Carson Woo	Sauder School of Business, Canada
Iryna Zolotaryova	Kharkiv National University of Economics, Ukraine
Joze Zupancic	University of Maribor, Slovenia

Organizing Committee

Chair

Stanislaw Wrycza

Secretary

Anna Węsierska	Department of Business Informatics at University of Gdansk

Members

Dorota Buchnowska	Department of Business Informatics at University of Gdansk
Bartłomiej Gawin	Department of Business Informatics at University of Gdansk
Przemyslaw Jatkiewicz	Department of Business Informatics at University of Gdansk
Dariusz Kralewski	Department of Business Informatics at University of Gdansk
Michał Kuciapski	Department of Business Informatics at University of Gdansk
Bartosz Marcinkowski	Department of Business Informatics at University of Gdansk
Jacek Maslankowski	Department of Business Informatics at University of Gdansk

EuroSymposium 2018 Topics

Agile Methods
Big Data, Business Analytics
Blockchain Technology and Applications
Business Informatics
Business Process Modeling
Case Studies in SAND
Cloud Computing
Cognitive Issues in SAND
Conceptual Modeling
Crowdsourcing and Crowdfunding Models
Design Theory
Digital Services and Social Media
Enterprise Architecture
Enterprise Social Networks
ERP and CRM Systems
Ethical and Human Aspects of IS Development
Ethnographic, Anthropological Action and Entrepreneurial Research
Evolution of the IS Discipline
Human-Computer Interaction
Industry 4.0
Information Systems Development: Methodologies, Methods, Techniques and Tools
Internet of Things
Machine Learning

Model-Driven Architecture
New Paradigms, Formalisms, Approaches, Frameworks and Challenges in IS & SAND
Ontological Foundations and Intelligent Systems of SAND
Open Source Software (OSS) Solutions
Project Management
Quality Assurance in Systems Development
Requirements and Software Engineering
Research Methodologies in SAND
Role of SAND in Mobile and Internet Applications Development
SAND Education: Curricula, E-learning, MOOCs, and Teaching Cases
SCRUM Approach
Security and Privacy Issues in IS and SAND
Service-Oriented Systems Development
Social Networking Services
Socio-Technical Approaches to System Development, Psychological and Behavioural
    Descriptions
Software Intensive Systems and Services
Strategic Information Systems in Enterprises
Supply Chain Management Aspects
Systems Analysts and Designers
Teams and Teamwork in IS & SAND
UML, SysML, BPMN
User Experience (UX) Design
Workflow Management

Contents

Healthcare IT

Systems Development and Engineering

A Performance Measurement System for Software Testing Process

Vuk Vukovic, Pere Tumbas, Lazar Rakovic,
Mirjana Maric, and Veselin Pavlicevic

Faculty of Economics in Subotica, University of Novi Sad,
Segedinski put 9-11, 24000 Subotica, Serbia
{vuk,ptumbas,lazar.rakovic,maricm,
pavlicevic}@ef.uns.ac.rs

Abstract. Proper management of development processes in software engineering can be achieved by means of continuous planning, measurement, monitoring and assessment of process performance indicators. The software testing process is also characterised by numerous, well-known performance indicators, based on which it is possible to plan and measure its performance. A question imposes itself, however, to which extent these performance indicators are really applicable in practice (measurable), sensitive and relevant to the software testing process. Theoretical and empirical research was conducted in order to obtain an answer to this research question. Experts in the software testing domain were surveyed, so that they would evaluate individual performance indicators of the software testing process, previously identified in papers published in referent scientific journals and conference proceedings. The result of conducted research in this paper is a developed performance measurement system for software testing process, with the primary objective to harmonise the existing differences between theory and practice in performance management of this process.

Keywords: Performance measurement system · Performance indicators
Software testing

1 Introduction

According to Stainer [1], performance can be described as "an organization's ability to achieve its goals". "Performance management implies systematic and integrated planning, channelling, coordinating and controlling activities in an organization for the purpose of effective and efficient achievement of set goals." [2]. For the organization to properly manage performance, it is inevitable to manage it at the level of its own processes [3]. The main prerequisite for successful performance management is the appropriate performance measurement system.

Historically, software development processes evolved for the purpose of more effective and efficient development of the software product [4]. With its verification and validation activities, software testing is one of the key points for securing the quality of

© Springer Nature Switzerland AG 2018
S. Wrycza and J. Maślankowski (Eds.): SIGSAND/PLAIS 2018, LNBIP 333, pp. 3–20, 2018.
https://doi.org/10.1007/978-3-030-00060-8_1

software products [5], why it is highly important for its performance to be at a high level.

The primary problem when planning and managing the software testing process is manifested in insufficiently systematic approach, unlike the software development process, where this problem is approached appropriately. Considering other dimensions of the testing process as well, rather than only costs and resource distribution, the software testing process can be optimized and better understood by the team manager and members [6].

Software testing performance measurement system should provide organizations involved in software production with a proactive position for enhancing its sub-processes and activities. Bearing this in mind, the aim of this paper is to define a performance measurement system for software testing process. It should create pre-conditions for better understanding, progress monitoring, evaluating and taking timely responses to unwanted deviations when realizing the software testing process.

In addition to the introduction, the paper consists of four sections. Section 2 presents research methodology of theoretical and empirical research. Section 3 contains the results and discussion, presenting the proposed performance measurement system for software testing process as the main result of the study and some limitations of the conducted study. Section 4 contains conclusions and avenues of future research.

2   Research Methodology

The selected methodology for achieving the research objective is presented in Fig. 1.

2.1   The Theoretical Concept for Method Definition

According to Balaban and Ristić [2], performance management process comprises three stages:

1. Performance planning;
2. Performance monitoring, measurement and assessment; and
3. Continuous performance improvement.

Defining the performance measurement system for software testing process is conceived at the first stage, performance planning, which, according to Balaban and Ristić [2] and Sommerville and Ransom [8] includes the following activities:

- Defining the expected performance (defining objectives, i.e. results/behaviours related to the organization's objectives);
- Identifying critical success factors (CSFs); and
- Determining performance indicators.

For the expected performance to be appropriately defined, it is primarily necessary to view and understand the set objectives and then define the CSFs of the software testing process. Definition of objectives and identification of CSFs of the software testing process was performed by the analysis of state of the art software testing models

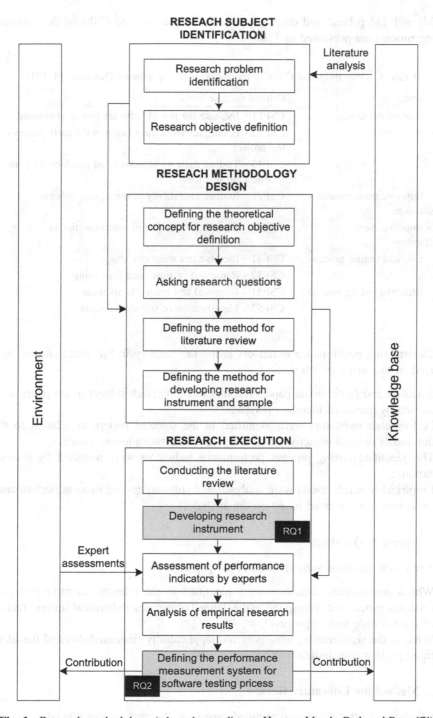

Fig. 1. Research methodology (adapted according to Hevner, March, Park and Ram [7])

TMMi and TMap Next and other sources. The objectives and CSFs of the software testing process are presented in Table 1.

Table 1. Objectives and CSFs of the software testing process (Sources: [8–12])

Objective	Critical success factor
O1- Increase the testing process maturity	CSF11 - Increase the use of software testing standards
	CSF12 - Testing activities must begin in the early phases of the project
	CSF13 - Testing must be objective and conducted by an independent test team
O2 - Improve management satisfaction	CSF21 - Reduce uncertainty in the testing process
O3 - Improve client satisfaction	CSF31 - Reduce the number of user objections
O4 - Efficient testing process	CSF41 - Increase test team efficiency
	CSF42 - Reduction of time used for testing
O5 - Effective testing process	CSF51 - Increased test team effectiveness
	CSF52 - Identification of software defects

Determining performance indicators is the key activity in this research, and was realized in four steps as follows:

- Scientific and professional papers relevant to the research subject in the paper were chosen by means of literature review;
- Performance indicators were identified in the selected papers in relation to the previously defined objectives and CSFs of the software testing process;
- The identified testing process performance indicators were assessed by domain experts;
- Empirical research results were analysed and software testing process performance indicators were selected based on the analysis.

2.2 Research Questions

Two research questions were defined in the paper:

- Which performance indicators were identified in the referent scientific and professional papers that correspond to the defined objectives and critical success factors of the software testing process?
- What is the significance, sensitivity and applicability (measurability) of the identified performance indicators?

2.3 Method for Literature Review

The theoretical research was performed based on the part of the procedure of the systematic literature review according to Kitchenham [13] and Kitchenham et al. [14].

The aim of the theoretical research review was to create prerequisites for obtaining the answer to RQ1. The formal search strategy encompassed the study material relevant to obtaining answers to RQ1. This strategy entails defining the research sources, electronic databases, and key words for search. The primary study material was taken over from Web of Science, Scopus and SpringerLink electronic databases.

The descriptors according to which the study material search was conducted are: "software testing metrics", "metrics in software testing", "metrics for software testing" "software testing process evaluation", "performance indicators" AND "software testing".

The criteria for including the study material were:

- Time dimension: refers to the search of study material published from 2000 till 2017;
- Type of study: study material matching the defined search descriptors in the title and/or the abstract. The study material refers to peer-reviewed scientific and professional articles published in scientific journals and proceedings from workshops and scientific conferences.

Criteria for exclusion of study material: Papers whose contents do not include software testing process metrics/indicators.

2.4 Method, Instrument and Sample for Empirical Research

The list of software testing process performance indicators (a data gathering instrument) was created according to the methodology described in Sect. 2.1. It contains objectives, CSFs and software testing process performance indicators (Sect. 3.1, Table 4). It was forwarded to respondents in the form of a check list, so that they could evaluate the identified software testing process performance indicators according to the following criteria:

- Criterion 1: significance of performance indicators for attaining a certain goal (1 – marginally significant, 2 – significant, 3 – very significant);
- Criterion 2: sensitivity (discriminativeness) of performance indicators – power to detect even slight changes in performance (1 – insufficiently sensitive, 2 – sensitive, 3 – very sensitive);
- Criterion 3: applicability of the performance indicator – to that extent is the performance indicator applicable in practice (0 – not applicable, 1 – applicable).

In addition to predefined assessments, the respondents were able to comment their own assessment of each software testing process performance indicator. After the analysis of respondents' assessments, certain performance indicators were included, and some were excluded from the software testing process performance measurement system.

The respondent sample comprised 20 domain experts from various software organizations on the territory of the Republic of Serbia. Criteria for selection of respondents included: several years of experience (minimum 5 years) and involvement in the tasks (role in the process) of software testing: test team director, test manager or QA lead.

3   Results and Discussion

This section gives answers to the posed research questions through the prism or results of the performed literature review and empirical research.

3.1   Which Performance Indicators Were Identified in Referent Scientific and Professional Papers that Correspond to the Defined Objectives and Critical Success Factors of Software Testing Process?

The study material search was conducted within defined electronic databases, according to the descriptors. The overview of the total number of hits, according to descriptors and stages of analysis for every electronic database, is given in Table 2.

Table 2. Search results of electronic databases

Data sources	Descriptors	Number of hits by descriptors (1st stage)	Number of papers included in further analysis (2nd stage)	Number of papers included in quality evaluation (3rd stage)
Web of Sci.	software testing metrics	9	3	3
Web of Sci.	metrics in software testing	2	1	1
Web of Sci.	metrics for software testing	2	1	1
Web of Sci.	software testing process evaluation	1	0	0
Web of Sci.	performance indicators AND software testing	2	0	0
SpringerLink	software testing metrics	24	3	2
SpringerLink	metrics in software testing	3	1	0
SpringerLink	metrics for software testing	10	1	1
SpringerLink	software testing process evaluation	6	0	0
SpringerLink	performance indicators AND software testing	0	0	0
Scopus	software testing metrics	16	1	1
Scopus	metrics in software testing	3	0	0
Scopus	metrics for software testing	6	1	0
Scopus	software testing process evaluation	0	0	0
Scopus	performance indicators AND software testing	44	0	0
Total		128	12	9

The first stage of the analysis of obtained data was focused on the analysis of titles and abstract. Metadata on papers (title, author, source and abstract) related to the defined research question by title and abstract were saved in an Excel spreadsheet.

In the second stage, the previously made Excel spreadsheet was analyzed, with the aim of identified papers meeting the criteria for exclusion of scientific material.

In the third stage, the papers which have duplicates in other electronic databases were excluded.

The process of detailed analysis included 8 papers in their entirety, in order to apply some of the defined quality criteria for their evaluation according to Dyba and Dingsoyr [15]. Having completed quality evaluation, it was established that all selected papers (9) met the quality criteria (Table 3).

Table 3. Papers meeting the literature review criteria

ID	Authors	Year	Title of the paper	Type of source
01	Kan, S. H., Parrish, J., & Manlove, D.	2001	In-process metrics for software testing	Scientific Journal
02	Afzal, W., & Torkar, R.	2008	Incorporating Metrics in an Organizational Test Strategy	Conference proceedings
03	Nirpal, P. B., & Kale, K.	2011	A Brief Overview Of Software Testing Metrics	Scientific Journal
04	Singh, Y., Kaur, A., & Suri, B.	2008	An Empirical Study of Product Metrics	Conference proceedings
05	Kanij, T., Merkel, R. & Grundy, J.	2012	Performance Assessment Metrics for Software Testers	Conference proceedings
06	Lazic, Lj., & Mastorakis, N.	2007	Cost Effective Software Test Metrics – Part 1	Conference proceedings
07	Lazic, Lj., & Mastorakis, N.	2007	Cost Effective Software Test Metrics – Part 2	Conference proceedings
08	David J. Barnes and Tim R. Hopkins	2006	Applying Software Testing Metrics to Lapack	Conference proceedings
09	Eldo, K. J. & Maheswari, D.	2015	Survey on software measurement systems based on software metrics	Scientific Journal

Table 4 shows performance indicators identified from the selected scientific papers. Their choice was determined by previously defined objectives and CSFs of the software testing process. They were further subjected to empirical evaluation by domain experts, the results of which are shown in the subsequent section.

Table 4. Performance indicator list according to objectives and CSF of the software testing process (Sources: [6, 8–12, 16–25])

Objective	Critical success factor	Performance indicator
O1 - Increase the testing process maturity	CSF11 - Increase the use of software testing standards	PI111 - Number of used testing standards
	CSF12 - Testing activities must begin in the early phases of the project	PI121 - Percentage of errors detected when testing (overviewing) information requirements in proportion to the total number of software defects
		PI122 - Percentage of errors detected when testing the design specification in proportion to the total number of software defects
	CSF13 - Testing must be objective and conducted by an independent test team	PI131 - Percentage of test cases executed by the independent test team
O2 - Improve management satisfaction	CSF21 - Reduce uncertainty in the testing process	PI211 - Percentage of successfully completed test cases in proportion to the total number of test cases defined in the testing plan
		PI212 - Coverage of testing requirements
		PI213 - Percentage of coverage of information (users') requirements by tests
O3 - Improve client satisfaction	CSF31 - Reduce the number of user objections	PI311 - Number of software defects detected in the maintenance period (after the release)
		PI312 - Number of software defects detected in user acceptance testing (UAT), before the release has been accepted
O4 - Efficient testing process	CSF41 - Increase test team efficiency	PI411 - Percentage of software defects defined by the software tester in proportion to the total number of identified software defects in the software product
		PI412 - Percentage of critical software defects identified by the software tester in proportion to the total number of software defects identified by the software tester
		PI413 - Number of executed test cases in an 8-hour working day per software tester
	CSF42 - Reduction of time used for testing	PI421 - Testing time (in hours) in proportion to the program code size (KLOC) of the software product
		PI422 - Testing time in proportion to development time of the software product
		PI423 - Proportion of manual and automated test cases in the software project
O5 - Effective testing process	CSF51 -Increased test team effectiveness	PI511 - Percentage of critical software defects in proportion to the total number of identified software defects in the software product
		PI512 - Percentage of discarded software defects in proportion to the total number of identified software defects in the software product

(continued)

Table 4. (*continued*)

Objective	Critical success factor	Performance indicator
		PI513 - Number of test cases in proportion to the number of identified software defects in the software product
		PI514 - Percentage of test cases that identified software defects in proportion to the total number of test cases
	CSF52 - Identification of software defects	PI521 - Percentage of software defects identified in the testing process in proportion to the total number of software defects of the software product
		PI522 - Proportion of the number of software defect in the testing cycle (testing process phase) to the total number of software defects in the testing process
		PI523 - Total number of identified software defects in a working day
		PI524 - Cumulative sum of identified software defects in the software product by weeks
		PI525 - Number of identified software defects by functional areas (modules) of the software product
		PI526 - Proportion of identified software defects to the size of the program code
		PI527 - Percentage of software defects caused by performance tests in proportion to the total number of software defects in the software product
		PI528 - Total number of identified software defects in proportion to the total number of identified software errors of the previous release of the software product
		PI529 - Total number of critical software defects in proportion to the total number of identified critical software defects of the previous release

3.2 What Is the Significance, Sensitivity and Applicability of the Identified Performance Indicators?

The average value of respondents' assessments for performance indicators according to "significance" (column 2) and "sensitivity" (column 3) assessment criteria, and the number of positive values whereby the respondents assessed performance indicators according to "applicability" (column 4) are shown in Table 5. A comparative overview of assessments according to three assessment criteria will enable a clearer insight for the analysis of significance, sensitivity and applicability of performance indicators.

Table 5. Comparative overview of assessments according to assessment criteria

PI	Average value of the significance criterion	Average value of the sensitivity criterion	Number of positive assessments for the applicability criterion
PI111	2.00	2.00	18
PI121	2.60	2.30	16
PI122	2.40	2.00	18
PI131	2.10	2.10	18
PI211	2.00	2.10	16
PI212	2.60	2.30	14
PI213	2.80	2.70	16
PI311	2.40	2.20	18
PI312	2.40	2.20	20
PI411	2.00	**1.90**	**14**
PI412	2.20	2.20	18
PI413	**1.50**	**1.60**	**12**
PI421	**1.40**	**1.60**	**10**
PI422	**1.60**	**1.60**	**14**
PI423	2.20	2.00	18
PI511	2.50	2.40	18
PI512	2.30	2.30	20
PI513	**1.60**	**1.90**	16
PI514	**1.90**	**1.70**	16
PI521	2.60	2.30	20
PI522	**1.70**	**1.60**	20
PI523	**1.20**	**1.10**	**10**
PI524	**1.40**	**1.40**	**12**
PI525	2.10	2.00	18
PI526	**1.50**	**1.40**	**12**
PI527	**1.70**	**1.80**	16
PI528	**1.90**	**1.80**	20
PI529	2.30	2.10	20

Based on Table 5, "lower" and "higher" assessed performance indicators were identified according to three assessment criteria in the conducted empirical research. Performance indicators with average assessment values lower than 2, average assessment value according to sensitivity criterion lower than 2, and value of the number of positive assessments for applicability criterion lower than or equal to 14 were identified as lower assessed. As for applicability criterion, the threshold was set high, due to the intention to identify indicators that have high applicability in practice.

Based on the set standards from the list of software testing process performance indicators, the lowest accessed performance indicators were excluded in the first iteration according to all three assessment criteria with markers:

- PI413 - Number of executed test cases in an 8-hour working day per software tester;
- PI421 - Testing time (in hours) in proportion to the program code size (KLOC) of the software product;
- PI422 - Testing time in proportion to development time of the software product;
- PI523 - Total number of identified software defects in a working day;
- PI524 - Cumulative sum of identified software defects in the software product by weeks and
- PI526 - Proportion of identified software defects in proportion to the size of the program code.

Performance indicators with values lower than 2 according to significance and sensitivity assessment criteria were excluded from the list of software testing process performance indicators in the second iteration:

- PI513 - Number of test cases in proportion to the number of identified software defects in the software product;
- PI514 - Percentage of test cases that identified software defects in proportion to the total number of test cases;
- PI522 - Proportion of the number of software defects in the testing cycle (testing process phase) to the total number of software defects in the testing process;
- PI527- Percentage of software defects caused by performance tests in proportion to the total number of software defects in the software product;
- PI528 - Total number of identified software defects in proportion to the total number of identified software errors of the previous release of the software product.

A conclusion can be drawn that these performance indicators, despite their satisfactory applicability, have disputable significance and sensitivity for the testing process, and were therefore excluded from the list of performance indicators.

The performance indicator "PI411—Percentage of software defects defined by the software tester in proportion to the total number of identified software defects in the software product" was excluded from the list of software testing process performance indicators in the third iteration. This performance indicator was near the borderline for exclusion from the list of performance indicators according to the set standard, so an analysis of the qualitative component of this part of research was performed. It consisted of assigning assessor's comments to individual performance indicators. The PI411 performance indicator received two non-affirmative comments by two assessors, which practically brought about the decision that this indicator be excluded from the list of software testing performance indicators.

Performance indicators that, in accordance with the domain experts' assessments, met the set standard (criterion), became an integral part of the software testing process performance measurement system (Table 6). In addition to the testing process objective and CSFs to which they belong, Table 6 shows the method and frequency of measurement, as well as sources of data for individual performance indicators.

In the qualitative component of the research, each performance indicator received individual comment from assessors, the content of which may be useful when implementing the method in software development organizations. Despite having received a satisfactory number of positive assessments in terms of applicability (16 out of 20), the performance indicator "PI121 - Percentage of errors detected when testing (overviewing) information requirements in proportion to the total number of software defects" received, by contrast, comments by two assessors stating that it was hard to apply, in terms of difficulty of measurement, and of various types of resistance in development teams when it came to testing information requirements. In performance indicator "PI311 - Number of software defects detected in the maintenance period (after the release)", it is necessary to follow the software defects at the level of severity. A differentiation must be made between trivial and critical software defects when following this indicator. The performance indicator "PI412 (new id PI411) - Percentage of critical software defects identified by the software tester in proportion to the total number of software defects identified by the software tester" can also be worded as follows: "Number of critical software defects that the software tester did not detect and were detected afterwards". The "PI511 - Percentage of critical software defects in proportion to the total number of identified software defects in the software product" and "PI512 - Percentage of discarded software defects in proportion to the total number of identified software defects in the software product" performance indicator is characterised by the test team's focus on the most important things in the software product. The "PI525 (new id PI522) - Number of identified software defects by functional areas (modules) of the software product" performance indicator represents a prerequisite for a useful analysis of a software product's quality. The "PI529 (new id PI523) - Total number of critical software defect in proportion to the total number of identified critical software defects of the previous release" performance indicators must include a note that much attention should be paid to the complexity of changes between the two releases of the software product.

3.3   Threats to Validity

The level of agreement, expressed by Cohen's kappa coefficient, between domain experts (assessors) who participated in the assessment of performance indicators for the needs of defining a performance measurement system for software testing, is very low in most cases between individual assessors, bearing in mind recommendations for interpreting the kappa coefficient according to Landis and Koch [26]. An exception is 14 pairs of assessors whose value of kappa coefficient is in the interval from 0.41 to 0.60, which is, according to Landis and Koch [26], interpreted as medium agreement between assessors.

The quality of the utilized approach for the assessment of the defined performance indicators using the average value of the responses measured on the 3-point scale, as a criterion, could be raised on the higher level using some of the multi criteria decision analysis techniques.

3.4   A Performance Measurement System for Software Testing Process

A performance measurement system for software testing process, resulting from the conducted dual research, is shown in Table 6. It defines the objectives of the software testing process, CSFs for attaining these objectives, and also performance indicators, method, frequency and data sources for their measurement.

Table 6. Performance measurement system for software testing process

ID	Performance indicator definition	Measurement method/measure	Frequency of measurement	Data source
O1- Increase the testing process maturity				
CSF11 - Increase the use of software testing standards				
PI111	Number of used testing standards	$S_{tn} = \sum S_i$ S_{tn} – Total number of used testing standards S – Testing standard	At the level of the software product	Software project specification
CSF12 - Testing activities must begin in the early phases of the project				
PI121	Percentage of errors detected when testing (overviewing) information requirements in proportion to the total number of software defects	$P_{ir} = (E_{ir}/D_{tn}) * 100$ P_{ir} – Percentage of errors detected when testing (overviewing) information requirements in proportion to the total number of software defects E_{ir} – Number of errors detected when testing (overviewing) information requirements D_{tn} – Total number of software defects in the software product	At the level of the software product	Requirement specification, reports from overviews and inspections of information requirements
PI122	Percentage of errors detected when testing the design specification in proportion to the total number of software defects	$P_{ds} = (E_{ds}/D_{tn}) * 100$ P_{ds} – Percentage of errors detected when testing the design specification in proportion to the total number of software defects E_{ds} – Number of errors detected when testing (overviewing) the design specification D_{tn} – Total number of software defects in the software product	At the level of the software product	Design specification, reports from overviews and inspections of design specifications
CSF13 - Testing must be objective and conducted by an independent test team				
PI131	Percentage of test cases executed by the independent test team	$P_{itt} = (T_{itt}/T_{tn}) * 100$ P_{itt} – Percentage of test cases executed by the independent test team T_{ntt} – Number of test cases executed by the independent test team T_{tn} – Total number of executed test cases	At the level of the software product	Test management system
O2 - Improve management satisfaction				
CSF21 - Reduce uncertainty in the testing process				
PI211	Percentage of successfully completed test cases in	$P_{sctc} = (T_{sctc}/T_{tnptc}) * 100$	At the testing cycle (stage)	Test management system

(continued)

Table 6. (*continued*)

ID	Performance indicator definition	Measurement method/measure	Frequency of measurement	Data source
	proportion to the total number of test cases defined in the testing plan	P_{sctc} – Percentage of successfully completed test cases in proportion to the total number of test cases defined in the testing plan T_{sctc} – Number of successfully completed test cases T_{tnptc} – Total number of planned test cases	level or in the desired time period	
PI212	Coverage of testing requirements	Whether the testing request is performed or not. Mark every testing request with yes or no	At the testing cycle (stage) level or at the testing process level	Testing strategy, testing plan, test management system
PI213	Percentage of coverage of information (users') requirements by tests	$P_{cirt} = (I_{cirt}/I_{tnir}) * 100$ P_{cirt} – Percentage of coverage of information (users') requirements by tests I_{cirt} – Number of information (users') requirements covered by tests I_{tnir} – Total number of information (users') requirements	At the level of the software product	Coverage matrix of information requirements with tests, test management system

O3 - Improve client (user) satisfaction

CSF31 - Reduce the number of user objections

PI311	Number of software defects detected in the maintenance period (after the release)	$D_m = \sum D_i$ D_m – Number of software defects detected in the maintenance period D – Software defect detected in the maintenance period	At the level of the software product	Users, system for reporting software defects by users
PI312	Number of software defects detected in user acceptance testing, before the release has been accepted	$D_{uat} = \sum G_i$ D_{uat} – Number of software defects detected in user acceptance testing D – Software defect detected in the user acceptance testing phase	At the level of the software product	Report on identified software defects, test management system

O4 - Efficient testing process

CSF41 - Critical success factor: Increase test team efficiency

PI411[a]	Percentage of critical software defects identified by the software tester in	$P_{cdst} = (D_{cdsp}/D_{tnd}) * 100$ P_{cdst} – Percentage of critical software defects	At the level of the software product	Test management system

(*continued*)

Table 6. (*continued*)

ID	Performance indicator definition	Measurement method/measure	Frequency of measurement	Data source
	proportion to the total number of software defects identified by the software tester Note: The indicator can also be shown for other categories of software defects (minor, medium, major)	identified by the software tester in proportion to the total number of software defects identified by the software tester D_{cdsp} – Number of critical software defects in the software product identified by the software tester D_{tnd} – Total number of software defects in the software product identified by the software tester		
CSF42 - Reduction of time used for testing				
PI421[b]	Proportion of manual and automated test cases in the software project	T_m/T_a T_m – Number of manual test cases in the software project T_a – Number of automated test cases in the software project	At the level of the software product	Test management system, tools in which test cases are automated
O5 - Effective testing process				
CSF51 - Increased test team effectiveness				
PI511	Percentage of critical software defects in proportion to the total number of identified software defects in the software product Note: The indicator can also be shown for other categories of software defects (minor, medium, major)	$P_{cd} = (D_{cd}/D_{tnd}) * 100$ P_{cd} – Percentage of critical software defects in proportion to the total number of identified software defects in the software product D_{cd} – Number of critical software defects in the software product D_{tnd} – Total number of software defects in the software product	At the level of the software product	Test management system
PI512	Percentage of discarded software defects in proportion to the total number of identified software defects in the software product Note: A rejected software defect is a feature of software product "incorrectly" identified by the software tester	$P_{dd} = (D_{dd}/D_{tnd}) * 100$ P_{dd} – Percentage of discarded software defects D_{dd} – Number of discarded software defects D_{tnd} – Total number of software defects in the software product	At the level of the software product	Test management system

(*continued*)

Table 6. (*continued*)

ID	Performance indicator definition	Measurement method/measure	Frequency of measurement	Data source
CSF52 - Identification of software defects				
PI521	Percentage of software defects identified in the testing process in proportion to the total number of software defects of the software product	$P_{dtp} = (D_{dtp}/D_{tnd}) * 100$ P_{dtp} – Percentage of software defects identified in the testing process in proportion to the total number of software defects of the software product D_{dtp} – Number of software defects identified in the testing process D_{tnd} – Total number of software defects in the software product (sum of number of software defects identified in the testing process and number of software defects identified in the maintenance period)	At the level of the software product	Test management system, system for reporting software defects by users
PI522[c]	Number of identified software defects by functional areas (modules) of the software product Note: Compare the values of different functional areas	$D_{dfa} = \sum D_i$ D_{dfa} – Number of identified software defects by functional areas (modules) of the software product D – Software defect identified in the functional area (module) of the software product	At the level of the software product	Test management system
PI523[d]	Total number of critical software defects in proportion to the total number of identified critical software defects of the previous release	D_{tncd}/D_{tncdpr} D_{tncd} – Total number of critical software defects in the software product D_{tncdpr} – total number of identified critical software defects of the previous release of the software product	At the level of the software product	Test management system

[a]The ID of performance indicator in the research instrument was PI412
[b]The ID of performance indicator in the research instrument was PI423
[c]The ID of performance indicator in the research instrument was PI525
[d]The ID of performance indicator in the research instrument was PI529

4 Conclusions and Future Research

Performance management of the software testing process appears to be an area where insufficient empirical research has been conducted. Bearing that in mind, this paper proposes one of the possible performance measurement systems for software testing process. It was created as a result of dual, theoretical and empirical research where objectives, CSFs and testing process performance indicators were first identified from the selected scientific and professional literature, and then forwarded to domain experts for assessment. The gathered, and then processed results of the empirical research determined the architecture (content) of the software testing process performance measurement system, whereby an empirical component was added to a theoretical concept of performance planning. The empirical evaluation of performance indicators of the software testing process was aimed at incorporating into the method solely applicative performance indicators significant for achieving the objectives of the software testing process, capable of detecting even the slightest changes in the performance of the software testing process.

The research result, a performance measurement system for software testing process, is the source of important information and solid base for future research. Implementation of the proposed method in a number of software organizations, realising software testing at higher levels of TMMi, could also produce the following research questions: does the proposed performance measurement system for software testing process enable a realistic insight into the testing process performance in software organizations, and what are its empirical advantages and disadvantages.

References

1. Stainer, L.: Performance management and corporate social responsibility: the strategic connection. Strateg. Change 15(5), 253–264 (2006)
2. Balaban, N., Ristic, Z.: Upravljanje performansom. M&I Systems, Co., Novi Sad (2013)
3. Rummler, G., Brache, A.: Improving Performance, How to Manage the White Space on the Organization Chart. Jossey-Bass, San Francisco (1995)
4. Matkovic, P., Tumbas, P.: A comparative overview of the evolution of software development models. Int. J. Ind. Eng. Manag. (IJIEM) 1(4), 162–172 (2010)
5. Naik, K., Tripathy, P.: Software Testing and Quality Assurance: Theory and Practice. Wiley, New Jersey (2008)
6. Shimeall, S.C.: Managing the Testing Process (2002). http://www.stickyminds.com/sitewide.asp?ObjectId=3650&Function=e. Accessed 03 Mar 2016 from StickyMinds.com: Software Testing & QA Online Community
7. Hevner, A.R., March, S.T., Park, J., Ram, S.: Design science in information systems research. MIS Q. 28(1), 75–105 (2004)
8. Sommerville, I., Ransom, J.: An empirical study of industrial requirements engineering process assessment and improvement. ACM Trans. Softw. Eng. Methodol. 14(1), 85–117 (2005)
9. TMMi Foundation: Test Maturity Model integration (TMMi) v3.1. TMMi Foundation, Dublin (2010)

10. Koomen, T., van der Aalst, L., Broekman, B., Vroon, M.: TMap Next. UTN Publishers, Hertogenbosch (2006)
11. Strong QA: Testing Strategy (2016). https://strongqa.com/uploads/document/doc/41/testing-strategy.doc. Accessed 04 Mar 2016 sa Strong QA
12. Rational Team: Rational Unified Process, v7.0. IBM Rational Software Corporation, Armonk (2005)
13. Kitchenham, B.A.: Procedures for Performing Systematic Reviews. Keele University, Eversleigh (2004)
14. Kitchenham, B.A., Pearl Brereton, O., Budgen, D., Turner, M., Bailey, J., Linkman, S.: Systematic literature reviews in software engineering - a systematic literature review. Inf. Softw. Technol. **51**, 7–15 (2009)
15. Dyba, T., Dingsoyr, T.: Empirical studies of agile software development: a systematic review. Inf. Softw. Technol. **50**, 833–859 (2008)
16. Kan, S.H., Parrish, J., Manlove, D.: In-process metrics for software testing. IBM Syst. J. **40**(1), 220–241 (2001)
17. Barnes, D.J., Hopkins, T.R.: Applying software testing metrics to lapack. In: Dongarra, J., Madsen, K., Waśniewski, J. (eds.) PARA 2004. LNCS, vol. 3732, pp. 228–236. Springer, Heidelberg (2006). https://doi.org/10.1007/11558958_27
18. Nirpal, P.B., Kale, K.: A brief overview of software testing metrics. Int. J. Comput. Sci. Eng. IJCSE **3**(1), 204–2012 (2011)
19. Singh, Y., Kaur, A., Suri, B.: An empirical study of product metrics. In: Iskander, M. (ed.) Innovative Techniques in Instruction Technology, E-learning, E-assessment, and Education, pp. 64–72. Springer, London (2008). https://doi.org/10.1007/978-1-4020-8739-4_12
20. Afzal, W., Torkar, R.: Incorporating metrics in an organizational test strategy. In: International Conference on Software Testing Verification and Validation Workshop, pp. 304–315. IEEE Computer Society, Washington (2008)
21. Baroudi, R.: KPI Mega Library. Scotts Valley (2010)
22. Kanij, T., Merkel, R.G., Grundy, J.: Performance assessment metrics for software testers. In: Sharp, H., Dittrich, Y., de Souza, C.R.B., Cataldo, M., Hoda, R. (eds.) 2012 5th International Workshop on Co-operative and Human Aspects of Software Engineering (CHASE), Proceedings, pp. 63–65. IEEE, Institute of Electrical and Electronics Engineers, Piscataway (2012)
23. Lazic, L., Mastorakis, N.: Cost effective software test metrics – Part 2. In: Proceedings of the 4th IASME/WSEAS International Conference on Engineering Education, pp. 144–153 (2007)
24. Lazic, L., Mastorakis, N.: Cost effective software test metrics – Part 1. In: Proceedings of the 4th IASME/WSEAS International Conference on Engineering Education, pp. 137–143 (2007)
25. Eldo, K.J., Maheswari, D.: Survey on software measurement systems based on software metrics. Int. J. Appl. Eng. Res. **10**(19), 40090–40095 (2015)
26. Landis, J.R., Koch, G.G.: The measurement of observer agreement for categorical data. Biometrics **33**, 159–174 (1977)

Towards Improving the Search Quality on the Trading Platforms

Olga Cherednichenko, Maryna Vovk[✉], Olga Kanishcheva,
and Mikhail Godlevskyi

National Technical University "Kharkiv Polytechnic Institute", 2, Kyrpychova
str., Kharkiv 61002, Ukraine
olha.cherednichenko@gmail.com, marihavovk@gmail.com,
kanichshevaolga@gmail.com,
mikhail.godlevskij@gmail.com

Abstract. In this paper, the problem of the search quality on the trading platforms, such AliExpress, eBay and others is explored, the major types of problems that arise in product search by customers are considered. The usage of the classical clusterization algorithms for grouping similar products according to their descriptions is studied. A data set for experimenting consists of different items (smartphones) from e-shop eBay is developed. Each entity in this corpus photos and a product description are given. These texts are used for item comparing in order to perform similar groups or similar items. The results show that the k-means algorithm is good for preliminary grouping but for detailed processing, other methods and approaches are required.

Keywords: Trading platform · Recommendation system · Product search
Information Technology · Product Classification

1 Introduction

Electronic-commerce (e-commerce) has become an important channel for business performing. The share of e-commerce is increasing continuously (Table 1). According to [1] the Internet retail becomes a popular option for consumers.

Table 1. The share of Internet commerce in the retail trade in the world [1]

Year	2008	2009	2010	2011	2012	2013	2014	2015	2016
Share of Internet commerce, %	4	4.4	5	5.7	6.5	7.2	7.9	8.6	9.3

Over the last decade the purchase process has been changed drastically. The first e-commerce shops were similar to real shops. Customers used to choose type and model of commodity from the finite set of goods. The interface and installation-specific settings were adjusted for that purpose. The appearance of huge trading platforms, like

S. Wrycza and J. Maślankowski (Eds.): SIGSAND/PLAIS 2018, LNBIP 333, pp. 21–30, 2018.
https://doi.org/10.1007/978-3-030-00060-8_2

AliExpress, eBay, Amazon etc., has changed the retail process. As a rule, trading platforms combine a huge number of sellers and goods. There are many options and product alternatives from different suppliers on different trading platforms. The set of available products is typically huge, it changes constantly, and new items are added. In such circumstances, a customer should choose where, what and from whom to buy.

Thus shoppers should look through plenty of pages in order to find an appropriate product. Finding the most advantageous offer for online shoppers has to provide different seller offers, compare product descriptions and images. As a result, the search space enhances dramatically. A lot of products cannot be sold when customers are not able to find them. It's obvious, that sellers are willing to promote their goods. They adjust to the recommended algorithms, which are used on certain trading platforms. In order to be present at as many search results as possible, sellers intentionally change the name of the commodity, photos and item's characteristics of the item. Thus the problem of search quality on the trading platforms is crucial.

To improve the process of product search, we look into methods from machine learning. In order to simplify the shopper's search, it is necessary to form similar products in groups. It would be useful to have an algorithm that can compare items and define the referenced commodity which fits best to the item group in order to make the search process easier, faster and more precise. For example, such algorithm is implemented in trading platform eBay. Choosing option "similar product" the groups of similar goods can be generated. The main drawback is low accuracy of those groups.

A search engine is a type of an information retrieval system which helps find the information stored in a computer system. All of the modern trading platforms provide the search engine in order to help shoppers. Recommendation systems also try to improve buying process by predicting what items are interesting and useful to the buyer based on specific information about the user profile or product information. However, nowadays recommendation systems and search engine on modern trading platforms are not able to solve all problems related to the quality search, such as incorrect product description, fake photos, and non-relevant search output.

The aim of this paper is the analysis of main problems for product searching on different trading platforms and experimenting with machine learning algorithms in order to group the similar items and reduce search space for shoppers.

The rest of the paper is organized as follows: Sect. 2 studies related works and summarizes different methods for search quality on the trading platforms. In this section, we also discuss main problems for search quality. In Sect. 3 we describe our data set (test corpus) and clustering results with different entities from phone product category for reducing search space. Finally, in Sect. 4 we briefly sketch future work and present the conclusions.

2   Related Works and Background

2.1   Analysis of Relevant Works

There are a lot of studies which are dedicated to the issues of improving online shopping experiences for consumers. On the e-commerce trading platforms, where the number of choices is overwhelming, there is a need to filter, prioritize and efficiently deliver relevant information in order to alleviate the problem of information overload.

Such problem belongs to the tasks of information retrieval. Some researchers solve this issue using recommendation systems. The recommendation system is defined as a decision making strategy for users under complex information environments [2]. Papers [3–5] are devoted to studying, comparing and analyzing personalized recommendation systems. The main types of recommendation systems are distinguished: collaborative filtering, knowledge-based, effect-based, rule-based, and content-based recommendation systems. As it was demonstrated [4], each of them has some disadvantages, and it is concluded that the combined application of a variety of techniques should satisfy the actual needs better.

The paper [6] proposes an alternative approach to retrieve information from a given e-commerce website, by collecting data from the site's structure, retrieving semantic information in predefined locations and analyzing user's access logs. It gives the opportunity to predict users' future behaviour. Some researchers [7, 8] suggest the extension of the technology acceptance model for its application in the e-commerce field by adding four criterion variables, namely, purchase, access number, access total time, and access average time.

The authors in the work [9] used genetic algorithms to optimize a specified objective function related to a clustering task for search. They have done some experiments on synthetic and real-life data sets which show the utility of the proposed method. Their analysis of the results of the experiment shows that the proposed method may improve the final output of k-means.

In the paper [10] authors created a new approach to a product recommendation. They investigated the possibility of using a hybrid recommend consisting of content based clustering and connections between clusters using collaborative filtering to make good product recommendations. The algorithm is tested on real products and purchase data from two different companies - a big online bookstore and a smaller online clothing store.

A lot of works use methods of machine learning, such as in the work [11] authors analyze products on shopping sites (Amazon and eBay). They use machine learning classifiers for grouping product descriptions. Also, they propose to use clustering techniques to detect taxonomy evolution.

Thus, we can conclude that many authors researched questions of data processing concerning goods on trading platforms. Different approaches were developed. In spite of that fact the formulation of the problem, which is given in our paper, isn't investigated. Our task is to research how groups of similar products can be distinguished based on item description in order to user could compare different products and then among similar products choose the best offer from different sellers.

2.2 Main Problems for Search Quality on Trading Platforms

There are two ways of search on trading platforms via keywords or item specifications. Most shoppers face the problem of the incorrect search result on their keyword request.

For example, on eBay site we choose the category Cell Phones & Smartphones, we use such filters as Format – Buy it now, Style – Bar, Condition – New and brand Samsung. At search results among Samsung models the following smartphone models as iPhone 7, iPhone 7 Plus LG Risio LG Treasure Sony Xperia XA ZTE Prestige are also presented. Apart from that covers for smartphones are also given at the search result.

Even using filters the search result may still contain errors. At filters, it is possible to choose Brand, Model, Color etc. But a seller also fills incorrect specification and we receive a lot of mistakes in a product description. While searching Apple iPhone Samsung Galaxy S8 is found. The seller accidently or intentionally put Brand – Apple and Model – Samsung Galaxy S8 (Fig. 1).

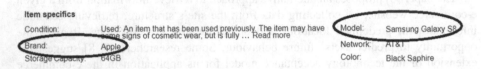

Fig. 1. Table of item specifics

Often, customers use photos for product search [12], but this does not greatly facilitate the search for the necessary product. Photos comparing is also a quite difficult task. Images of the same product can be presented in different ways. For example, smartphone's commodity picture can be presented in front or back side, be in the box or not, show the only photo or several (Fig. 2).

Fig. 2. Examples of different types of pictures

But all the mentioned approaches have some drawbacks. Using the recommendation systems requires considerable statistic data and doesn't allow comparing similar product items by their descriptions from different sellers. They provide propositions for buyers; however important information can be missing. User profiles in some cases can give additional information, but it is also problematic to classify users according to their type of behavior.

The most crucial thing is that on e-commerce trading platforms sellers try to influence on search system using words from the most top-selling product names. Because of that e-commerce user receives irrelevant search results (Fig. 1). There is also a problem of the significant difference between the product names of the same product in different sellers as well as inaccuracy and gaps in the item description.

3 Experiments

3.1 Data Set

Our task is to develop a service that will help customers quickly find the right product. For the experiment, the category of products "mobile phones" was chosen. Descriptions of goods are collected from different trading platforms. All found goods are entered into a single repository. The number of such commodity items is very large (for large trading platforms the number of offered phones reaches 370 thousand). This leads to the need to group goods. We assume that similar goods have similar descriptions. It is necessary to study how far the methods of machine learning can be used to group similar products on the basis of their descriptions. To collect and further process data, we have developed an architecture (Fig. 3). On the diagram, we can see that the designed system has 3 layers and 3 microservices, namely: Data representation layer, which contains classes and interfaces of data objects (data transfer object), classes representing the essence of the non-relational database and the pattern of design Repository, which consists of two data sources – a system cache and a non-relational Google Firebase database that works with cloud storage services.

The domain of the application layer is split from three microservices:

- agent platform;
- data seeker;
- data analyzer;
- parser and collector of data from Internet pages.

"Agent platform" service is responsible for executing business process processes in the body of individual agents.

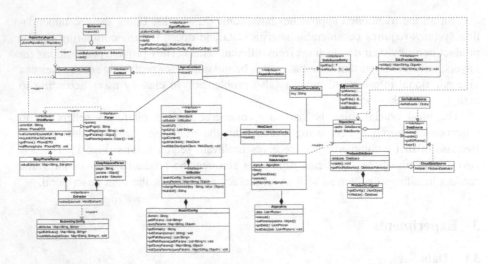

Fig. 3. Architecture of the software solution

The Data Finder service performs the process of searching for the required merchandise in trading venues by replacing the HTTP request parameters, filtering the configuration, and transitioning to data delivery pages.

The Parser tool analyzes an HTML document that answers an HTTP request from a searcher and collects the necessary data from the product description page.

The Data Analyzer service is responsible for analyzing the sampling collected by the parser, based on which the search parameters and product data stores will be reformed.

Due to the fact that the system has a clear division into layers of data representation and is distributed to separate microservices – testing processes are not labor-intensive. System components obviously do not depend on each other, the input data can be replaced by mock objects or test data for integrating and unit testing.

Thus, in order to solve the task of implementing the "Data Analyzer" service, it is necessary to choose an approach for putting products in order and constructing a model for estimating similarity. For this, a test set of product descriptions has been generated. The data is collected using the "Parser" service. The collected descriptions are stored as separate entities. For the experiment we use a sample of the description of smartphones from one site.

We have created own dataset from eBay.com website (https://www.ebay.com/). Our corpus contains 350 entities from Phone product category where each entity has 3–15 photos and product description. Some statistics are shown in Table 2. The example of product description and photos can be seen below (Fig. 4).

Name - SAMSUNG GALAXY S7 EDGE 32GB ORO SM-G935V VERIZON DESBLOQUEADO SMARTPHONE 5.5"
URL - https://www.ebay.com/itm/Samsung-Galaxy-S7-Edge-32GB-Oro-SM-G935V-Verizon-Desbloqueado-Smartphone-5-5/202155524056?epid=23011932310&hash=item2f11687fd8:g:9cMAAOSwaSZaO5fL
Condition - New: A brand-newunusedunopenedundamaged item in its original packaging (where packaging is applicable). Packaging should be the same as what is found in a retail storeunless the item is handmade or was packaged by the manufacturer in non-retail packagingsuch as an unprinted box or plastic bag. See the sellers listing for full details. See all condition definitions- opens in a new window or tab ... Read moreabout the condition
    MPN - SAM-G935V
    Cámara - 12 megapxeles
    Memoria RAM - 4 GB
    Memoria interna - 32 GB
    Tipo - Barra
    EAN - 0711202910423
    Marca - Samsung
    Modelo - Samsung Galaxy S7 Edge
    Tamaño de pantalla - 5,5"
    Color principal - Oro
    Procesador - Quad core
    Sistema operativo – Android

Fig. 4. The example of Smartphone photos with product description

Table 2. Statistics about data set

Category (Phone)	Number of entities
Galaxy S5	50
HTC Desire 816	50
iPhone 7	50
Nokia 1100	50
Sumsung Galaxy S7	50
Sony Xperia Z2	50
Sony Z5 Premium	50
Total	*350*

Such subcategories of product description as Name and Condition may contain very different texts. They may be different in length, words etc. because these sentences are written by the sellers. Other subcategories (Memory, Model, System etc.) are more similar in different descriptions. As a result the product description contains a lot of mistakes.

3.2   Experiments for Reducing Search Space

For the first stage, we try to use classical k-means clusterization algorithm for all product descriptions. Our algorithm uses the stopword list and TFIDF for vectorization of our texts. We received 350 samples with 2,659 features. Below can see the top terms per cluster (we take top 10 keywords for each cluster). Top terms per clusters are:

Cluster 0: used condition item apple 128gb iphone cosmetic fully functions previously
Cluster 1: sony xperia z5 premium z2 condition packaging e6853 new 32gb
Cluster 2: s7 samsung g930v galaxy 32gb edge sm g935v verizon unlocked
Cluster 3: htc desire 816 8gb 13mp sim dual condition android unlocked
Cluster 4: nokia 1100 apple iphone unlocked phone condition black gsm manufacturer
Cluster 5: s7 samsung 32gb galaxy sm unlocked smartphone 4gb condition lte
Cluster 6: samsung s5 galaxy 16gb 16mp packaging condition g900v 4g retail

As shown in the cluster list, three clusters (in blue color and underlined) have similar terms and are related to Samsung smartphone. However, we know that we only have two categories of Samsung brand: Galaxy S5 and Samsung Galaxy S7. This result is not very good for such a small sample, so we try to use the Porter stemmer for preprocessing our dataset. Our results with top terms per clusters are:

Cluster 0: nokia 1100 phone unlock mobil condit black manufactur cellular refurbish
Cluster 1: use condit item floor previous cosmet wearbut return fulli store
Cluster 2: s7 samsung galaxi 32gb sm unlock g930v packag smartphon 4gb
Cluster 3: samsung s5 galaxi 16gb packag 16mp condit retail sm g900v
Cluster 4: appl iphon 128gb memori unlock condit io 32gb built smartphon
Cluster 5: htc desir 816 8gb 13mp sim dual condit unlock android
Cluster 6: soni xperia z5 premium packag z2 condit e6853 new unlock

In this case, we received two clusters for Samsung smartphone but only one for Sony smartphone. One cluster (in blue color and underlined) does not have a certain category and contains general keywords. As our next step, we use HashingVectorizer which hashes word occurrences in a fixed dimensional space. The word count vectors are normalized to each have l2-norm equal to one (projected to the Euclidean unit-ball) which be important for k-means to work in high dimensional space.

HashingVectorizer does not provide IDF weighting as this is a stateless model. When IDF weighting is needed it can be added by pipelining its output to a TfidfTransformer instance.

It can be noted that k-means (and minibatch k-means) are very sensitive to feature scaling and that in this case, the IDF weighting helps improve the quality of the clustering by quite a lot as measured against the "ground truth" provided by the class label assignments of our dataset.

After all steps, k-means works with 837 features. Top terms per clusters:

- _Cluster 0:_ s7 samsung galaxy 32gb sm edge g930v smartphone 4 gb contract
- _Cluster 1:_ htc desire 816 8gb 13mp used sim dual android mobile
- _Cluster 2:_ sony xperia z2 d6503 z3 case cover retail compatible 16gb
- _Cluster 3:_ apple iphone 128gb memory used 32gb smartphone ios ohne black
- _Cluster 4:_ samsung s5 galaxy 16gb 16mp 4g retail g900v lte smartphone
- _Cluster 5:_ nokia 1100 phone mobile black germany network refurbished gsm used
- _Cluster 6:_ sony z5 premium xperia e6853 32gb 23mp smartphone 3gb black.

For this k-means version, we receive good results. We estimated the quality of received results for 350 samples with 2,659 features represented on a trading platform. All our entities (smartphones) have a separate category. We receive good small value for the Silhouette Coefficient (_Silhouette Coefficient: 0.125_) which is actually for high dimensional datasets such as text data. Other measures such as Precision, Recall are also very good, their values being 0.95, 0.95 respectively.

As an experiment, we try to see how the k-means algorithm works with incorrect data. For that, we take a product description which contains the word Samsung in the Name field, and the word Apple in the Brand field.

The experiment shows, the algorithm is very sensitive to such data. If the Name field contains any words offering to Samsung models (e.g., Samsung Galaxy S5), then the algorithm takes this product to the Samsung category. However if the Name field contains only keyword Samsung without any other words, then this description is taken to the category Apple.

Thus, the usage of the clustering algorithm does not allow identifying unscrupulous sellers. But it provides relevant clusters and gives a possibility to distinguish product items. The next step can be processing data in the separate group.

4   Conclusions and Future Works

In this paper, experimenting results of product item description preprocessing are presented in order to compare and define the similar items. It provides the ability to decrease search space and to build an algorithm for buyer assistance. The obtained results of this study can help in future to create the integrated method for disambiguation of item description in order to improve the search quality, whereas only clusterization method could not help with solutions to all problems.

However, the k-means method can be used for preprocessing of the item description. The creation of a combined method, using classical algorithms and specific approaches, will allow increasing of search quality and buyer satisfaction.

In future works, it is supposed to create an approach to commodity grouping, which will combine product description and photo. It'll give the opportunity for a buyer to find the same products in reduced search space with dissimilar pictures and names, to find the same sellers at different trading platforms. It is supposed to make experiments not only with smartphones but with clothes and bags also.

References

1. Internet trading in Ukraine. https://netpeak.net/ru/blog/15-slaydov-o-tom-kak-razvivaetsya-rynok-elektronnoy-kommercii-v-ukraine/. Accessed 29 Jan 2018
2. Rashid, A.M., et al.: Getting to know you: learning new user preferences in recommender systems. In: Proceedings of the International Conference on Intelligent User Interfaces, San Francisco, California, USA, pp. 127–134 (2002). https://doi.org/10.1145/502716.502737
3. Ya, L.: The comparison of personalization recommendation for e-commerce. In: International Conference on Solid State Devices and Materials Science, 1–2 April 2012, Macao, vol. 25, pp. 475–478 (2012). Physics Procedia
4. Isinkaye, F.O., Folajimi, Y.O., Ojokoh, B.A.: Recommendation systems: principles, methods and evaluation. Egypt. Inform. J. **16**, 261–273 (2015)
5. Ying, L., Boqin, L.: Application of transfer learning in task recommendation system. Procedia Eng. **174**, 518–523 (2017)
6. Dias, J.P., Ferreira, H.S.: Automating the extraction of static content and dynamic behaviour from e-Commerce websites. Procedia Comput. Sci. **109**, 297–304 (2017)
7. Fayad, R., Paper, D.: The technology acceptance model e-commerce extension: a conceptual framework. Procedia Econ. Finance **26**, 1000–1006 (2015)
8. Kumar Raja, D.R., Pushpa, S.: Feature level review table generation for E-Commerce websites to produce qualitative rating of the products. Future Comput. Inform. J. **2**(2), 118–124 (2017)
9. Murthya, C.A., Chowdhury, N.: In search of optimal clusters using genetic algorithms. Pattern Recogn. Lett. **17**(8), 825–832 (1996). https://doi.org/10.1016/0167-8655(96)00043-8
10. Hansson, L.: Product Recommendations in E-commerce Systems using Content-based Clustering and Collaborative Filtering. https://lup.lub.lu.se/student-papers/search/publication/7860347. Accessed 05 July 2018
11. Bankar, S., Anindya, D.: Clustering for Taxonomy Evolution. http://web.cs.iastate.edu/~sabankar/Clustering_Report.pdf. Accessed 05 July 2018
12. Kiapour, M.H., Han, X., Lazebnik, S., Berg, A.C., Berg, T.L.: Where to buy it: matching street clothing photos in online shops. In: 2015 IEEE International Conference on Computer Vision (ICCV), Santiago, Chile, pp. 3343–3351 (2015). https://doi.org/10.1109/iccv.2015.382

A Review of Fundamental Tasks in Requirements Elicitation

Ramandeep Kaur Sandhu$^{(\boxtimes)}$ and Heinz Roland Weistroffer

Virginia Commonwealth University, Richmond, USA
{Sandhurk2,hrweistr}@vcu.edu

Abstract. The success of a software system implementation is contingent on how well it meets the needs or requirements of the organization and the end users. Many information systems projects get abandoned before completion or fail to deliver the expected organizational benefits, causing significant losses in resources, productivity, and reputation. The main reason for the failure of these projects is inaccurate or incomplete requirements determination. Thus, the requirements elicitation process has a significant impact on information systems success and therefore is considered as the most critical workflow in the development process. The requirements elicitation process consists of several fundamental tasks and performing these tasks appropriately may contribute considerably towards a successful systems development effort. This study provides a comprehensive review of these fundamental tasks during the requirement elicitation process and the challenges associated with these tasks or activities.

Keywords: Requirement elicitation · Domain · Stakeholders · Techniques

1 Introduction

The success of a software system is contingent on how well it meets the needs or requirements of the organization and the end users [8]. Many information systems projects never reach completion or when completed, fail to deliver the expected organizational benefits, causing significant losses in resources, productivity, and reputation. The foremost reason for the failure of these projects is inaccurate or incomplete requirements determination. Thus, the requirements elicitation process is extremely important for information systems success and therefore is considered as the most critical workflow in the development process [7]. According to Zowghi and Coulin [35], the requirements elicitation process involves five fundamental activities: (a) identifying the application domain, (b) identifying the sources of requirements, (c) identifying and analyzing the stakeholders, (d) selecting techniques, approaches, and tools to use, and (e) eliciting the requirements from stakeholders and other sources.

Notably, published research has enumerated the fundamental tasks/activities that are carried out during the requirements elicitation process. Carrying out all the fundamental tasks appropriately holds much promise for enhancing the systems development effort, but there are many challenges associated with these tasks. Despite the acknowledgment of criticality of the requirements elicitation process, we are not aware

© Springer Nature Switzerland AG 2018
S. Wrycza and J. Maślankowski (Eds.): SIGSAND/PLAIS 2018, LNBIP 333, pp. 31–44, 2018.
https://doi.org/10.1007/978-3-030-00060-8_3

of any study that provides a comprehensive review of how these fundamental requirements elicitation tasks should be carried out and the challenges associated with these tasks.

Past reviews of the requirements elicitation process cursorily discuss the requirements elicitation tasks. The study by Sharma and Pandey [30] briefly presents the requirement elicitation process, issues, and challenges associated with different techniques used in requirement elicitation process. But the study doesn't provide a comprehensive review of how these fundamental tasks are carried out during the requirements elicitation process and the challenges associated with these tasks. Rather, the focus of the study is discussing the challenges associated with the techniques used in the requirements elicitation process. Zowghi and Coulin [35] provide a comprehensive review of techniques, approaches, and tools for the requirements elicitation process, briefly discussing five fundamental types of tasks that are carried out during requirements elicitation. However, the study doesn't provide any discussion on how these tasks are carried out or the challenges associated with these tasks and how the challenges can be overcome. Razail and Anwar [27] and Pacheco and Tovar [23] focus on selecting the right stakeholders' methods for requirements elicitation and the challenges associated with it. Fuentes-Fernández et al. [14] focus only on the human factors involved in the requirements elicitation process. Hadar et al. [17] study the positive or negative effects of the domain knowledge on requirements elicitation via interviews as perceived by analysts with or without domain knowledge.

The motivation for this study is the lack of a comprehensive review of how the fundamental types of activities are carried out during the requirements elicitation process and the challenges associated with these tasks. Our objective is to help system developers in assessing and mitigating risks and focusing on essential aspects of requirements elicitation to improve the likelihood of successful system development. Although there are many requirements elicitation activities listed in the literature, we limit our attention to only the five most fundamental types of requirements elicitation tasks suggested by Zowghi and Coulin [35].

The research questions we address in this study are:

RQ1: *How are the most fundamental tasks carried out during the requirements elicitation process?*
RQ2: *What are the challenges associated with carrying out these tasks?*
RQ3: *What are the best practices practitioners should adopt to successfully carry out these tasks and elicit accurate requirements for the proposed system?*

The contribution of our research is synthesizing the literature on the five fundamental types of requirements elicitation tasks/activities to make these more readily accessible to practitioners and future researchers. Additionally, our research will provide recommendation to practitioners for managing the challenges associated with these five fundamental types of requirements elicitation tasks, to make the requirements elicitation process sufficiently robust in order to gather the critical system requirements.

The remainder of this study is organized as follows: Sect. 2 describes the methodology adopted for this study. Section 3 provides the detailed explanation of the main findings of the study. Section 4 provides the discussion of the main finding and concludes the study.

2 Methodology

The methodology adopted for this study is a comprehensive literature review of studies published from 2006–2017. We selected qualitative, quantitative, and empirical studies published in academic journals and conferences using google scholar, IEEE Xplore, Elsevier Science Direct, and Springer Link. The search terms included fundamental types of requirement elicitation tasks listed by Zowghi and Coulin [35], challenges associated with these tasks, and requirements engineering (as requirements elicitation is one of the critical phases in the requirements engineering process). We didn't limit our research to information systems specific studies because we consider system development research as cross-disciplinary, expanding into the areas of management, computer science, accounting, healthcare administration, etc. A total of 70 articles were retrieved from our sources. Only studies we considered relevant to the topic and that were highly cited were included in our review. Initial relevancy was determined from the titles and abstracts of the articles. Studies considered relevant were then read in detail for further verification. Studies that don't contribute towards answering our research questions were excluded from the literature review. Thus, out of the originally 70 articles, a total of 35 studies were used for the review.

3 Fundamental Tasks in Requirements Elicitation

3.1 Understanding the Application Domain

Knowledge about the organization, i.e. its structure, business domain, objectives, and policies, knowledge about the domain, i.e. its fundamental concepts, objectives and regulations, and knowledge about the system-as-is are the three crucial areas to focus on while trying to gain an understanding of the application domain.

The knowledge acquisition process involves two parties: requirements analysts and stakeholders [12]. The requirements analysts may be domain expert or domain ignorant. Domain expert analysts may have acquired the domain knowledge by developing other systems within the same domain. Having prior domain knowledge helps with requirements-completeness as the analysts may better know what issues need to be covered and what issues can be left aside. Prior domain knowledge also permits the analysts to ask critical questions that are understandable by the stakeholders. On the other hand, domain ignorant analysts are likely to require extra effort in learning the basics of the domain and may not be aware of all the critical issues that need to be covered [17].

On the flip side, having in-depth domain knowledge may also lead to overlooking the obvious and making tacit assumptions about the domain, whereas lack of domain knowledge can promote ideas generation independent of any domain assumptions and lead to asking questions that reveal the issues that domain experts might have overlooked. Hence, a team consisting of a mix of domain experts and domain ignorant analysts can be useful in extracting the complete information [21].

The problem of ambiguity may also arise during communication between requirements analysts and stakeholders. Therefore, the analysts should be familiar with the nature of these ambiguities and should be provided with the cognitive tools to identify and mitigate them during the interview process. The analyst can use domain specific terms that are unknown to him/her as cues. In these situations, it becomes crucial for the analyst to identify when a common term is used with the particular meaning in the stakeholder's domain. For example, the term "program" may have a domain specific meaning to the stakeholder, unfamiliar to the analyst. Such cues can be used as reference to detect and mitigate the ambiguity [12].

Tacit knowledge is another challenge, which may lead to incomplete information about the application domain. Neither the intended system users nor the experts can always impart their knowledge accurately and succinctly in response to direct questions [10]. To overcome this challenge, the parties involved may undergo a *solitary requirement elicitation (SRE)* exercise, where the analysts and the stakeholders write down all the unspoken assumptions about the application domain and the expert knowledge that may be taken for granted. The domain knowledge of both parties is amalgamated together to gain the collective knowledge about the application domain [13]. Additionally, the exposure of analysts to the user's business environment can also help overcome the tacit knowledge challenge [33].

Studies related to mind and knowledge indicate that requirements analysts should take the initiative to instigate the stakeholders to divulge as much as they can. *"The information acts as the trigger that induces the selection of specific patterns in the mind. These impulses from outside are essential for 'actuating' knowledge or for arranging specific combinations of knowledge material into a specific structure and starting specific knowledge processes. It means that knowledge is context and outside impulse dependent"* [2]. It is imperative to provide the end users with the right impulses as well as the right context.

The procedures, challenges, and best practices related to "understanding the application domain" tasks are summarized in Table 1.

Table 1. Understanding the application domain

How	Challenges	Best practices the practitioners can adopt
Knowledge acquisition by the requirements analysts	Prior domain knowledge	Mix of domain expert and domain ignorant analysts [21]
	Ambiguity	Ambiguity awareness [12]
	Tacit knowledge	Solitary requirements elicitation exercise [13] Analyst involvement in targeted business environment [12, 33]
Knowledge delivery by the end users	Tacit knowledge	Solitary requirements elicitation (SRE) exercise [13] Knowledge triggering [2]

3.2   Identification and Analysis of the Key Stakeholders

Correct identification of the key stakeholders avoids requirements overlapping, improves requirements coverage, and allows for more rational organization of the requirements [23]. The key stakeholders are the individuals who

1. have vested interests in the system
2. must introduce, manage, or operate the system after its deployment
3. are involved in developing the system
4. are accountable for the processes the system automates
5. have financial responsibilities
6. constrain the system [16].

The stakeholders fall into four categories: primary, secondary, external, and extended. Primary stakeholders are individuals who have the authority, power, and responsibility over the financial resources. Secondary stakeholders are the individuals who are affected by the outcomes of the project indirectly. External stakeholders are the ones who add value to the project from outside but are not the part of the project. Extended stakeholders could be the individuals assisting primary and secondary stakeholders to reach their visions. The stakeholders vary throughout the system development life cycle [19], and not all stakeholders can be included in the project due to specific project constraints. Razali and Anwar [27] introduced a three-stage framework that helps in determining the key stakeholders: *identification, filtering, and prioritization.* The *identification* stage involves utilizing project goals, types, and domain to recognize the stakeholders' types and roles. The stakeholders' roles can be classified into three groups:

A. *Mandatory:* Stakeholders whose involvement in the system is mandatory, otherwise the system success will be threatened.
B. *Optional:* Stakeholders who are not essential, as disregarding their needs doesn't threaten the system success.
C. *Nice-to-have:* Stakeholders that don't much influence the success of the system.

Each role has a degree of importance. During *filtering*, the identified stakeholders are filtered by applying a stratified sampling method with priority given to the role with the higher degree of importance. Further, to make the selection process unbiased, some form of analysis based on knowledge and interest should be conducted. Methods such as personality testing or interest inventory can be adapted to determine the interest of the stakeholders and eligibility for prioritization.

Lastly, during the *prioritization* process, certain stakeholders are on the priority list based on their interpersonal skills, including negotiation, collaboration, and communication skills.

Identifying the key stakeholders may be more challenging in inter-organizational projects (IOPs) than in traditional ones. In IOPs, there is greater complexity due to greater simultaneous cooperation and competition among ION (inter-organizational networks) members. A five-step procedure introduced by Bellejos and Montanga [4] can overcome this challenge:

1. *Specifying stakeholder types:* The four criteria described below should be applied to build the profile characterization of stakeholders to be involved in the project.

   1.1 *Functional criterion:* Involves analyzing various processes that will be affected by inter-organizational information systems (IOSs), either directly or indirectly.

   1.2 *Geographical location criterion:* Involves analyzing various geographical areas that must be involved in the selection process.

   1.3 *Knowledge and abilities criterion:* Involves identifying specific knowledge and abilities about the process or the activities the IOS will support.

   1.4 *Hierarchical level criterion:* At the organizational level, every hierarchical level must be considered when specifying the stakeholder's types. There are diverse viewpoints and perspectives when distinct hierarchical levels of IONs (inter-organizational networks) are considered.

2. *Specifying the stakeholder roles:* This step determines the scope, characteristics, and participation of each role during the project.

3. *Selecting the concrete stakeholders:* This step guides the project team to select the concrete entities based on the conditions specified in step 1.

4. *Associating the stakeholders with roles:* This step requires associating the roles determined in step 2 to each stakeholder.

5. *Analyzing the stakeholder's influence and interest:* Involves analyzing the influence and interest of each identified stakeholder in the desired project by conducting a formal assessment.

The procedures, challenges, and best practices related to identification of key stakeholders are summarized in Table 2.

Table 2. Identification and analysis of key stakeholders

How	Challenges	Best practices the practitioners can adopt
All stages of project development	Project constraints	Selection based on • Identification • Filtering, and • Prioritization [27]
Involvement of all diverse organizations in IOPs	Complexity due to simultaneous cooperation and competition	Five-step procedure involving: 1. Specifying the stakeholder's types 2. Specifying stakeholder's roles 3. Selecting the stakeholders based on step 1 4. Associating the stakeholders with the roles based on step 2 5. Analyzing the stakeholder's influence and interest [4]

3.3 Identification of the Sources of Requirements Acquisition

The stakeholders, domain experts, existing system documentation, and existing systems are the key sources of requirements [22, 24, 28]. The factors such as information related to systems feature, innovative ideas, work context and workflows, organizational policies, standards, legislation, and market information, determine the source of requirements. The stakeholder representatives gather together for an intensely focused period to create and review the necessary requirements related to high-level features/innovative ideas for the new project. During these meetings, the analyst discusses the desired project/system with the stakeholders and establishes the understanding of the stakeholders' requirements. Since, the stakeholders play a variety of roles in distinct projects and their desires and goals are conformed to organizational business objectives only, they may lack the experience of system development and hence fail to provide the comprehensive information about the system features. In such cases, observing the existing systems can be an excellent source of requirements and can serve as the source for validating the existing information provided by the stakeholders during the conversational methods.

Observing the users of existing systems and making the detailed observation of all their cases becomes the main source of requirements when the information related to work flow/context and the initial understanding of the system is required [33]. The biggest concern is that the users may modify their behavior when being aware of being observed. To avoid this, the observer should try to participate in the users' daily activities with the minimal involvement in their decision-making [34]. While observing, the observer may also not be sure about which elements of the system need to be observed. Therefore, the observer should be provided appropriate system training before commencing any observational task.

The specifications of the legacy system and the reuse of the glossaries open up the domain requirements, organizational policies, standards, legislation, and the user interface requirements. The variety of the documents such as the problem analysis, standards, charts, user manuals of existing tasks, and survey reports of competitive markets, becomes the principal sources of the requirements. But if the individual analyzing these documents is not a domain expert, he/she may not be able to understand the technical terms, cause-effect relationships, and conceptual structure by just studying the documents. Hence, involvement of an expert, while studying the documents will help with clarification of the subjective and technical terms and the conceptual structure [33].

When acquiring information on market needs, online discussion forums and blogs can provide necessary information about new trends, modifications, or strategic behaviors [20]. Nowadays, information imbedded in the product may be retrieved and also serves as a good source for requirements [32]. However, the main concern is that the end-user feedback may be vast, diverse, and inconsistent. Therefore, more structured end-user feedback can help collecting meaningful requirements [20]. Also, a continuous adaptive requirement engineering framework that enables requirements engineering at runtime can serve as a good source for requirements [26].

However, there is weakness inherent in each source for requirements when used alone. Therefore, collecting information from different sources increases the credibility of the requirements [25].

The procedures, challenges, and best practices related to "identification of sources of requirements" task is summarized in Table 3.

Table 3. Identification of sources of requirements

How	Challenges	Best practices the practitioners can adopt
System features	Lack of system development experience	Combinations of conversational and observational methods [25]
Work process/flows	End users' behavior changes while being aware of being observed	Balance between observer role and involvement in user's community [33, 34]
	Lack of sufficient knowledge of the element that needs to be observed	Providing training on specific tasks [25]
Organizational policies, regulations, and standards	Understanding the technical terms, conceptual structure, and cause and effect relationships	Involvement of the domain expert while studying the documentation [34]
Market needs	Huge, diverse, and inconsistent online feedback	Structured end-user feedback [20]
		Self-adaptive systems capable of performing requirement engineering at runtime [26]

3.4   Selecting Techniques, Approaches, and Tools to Use

There are a variety of techniques, approaches and tools available, which can help elicit requirements. Most organizations choose multiple techniques, which vary from each other in terms of efficiency and quantity of information being gathered. Organizations select a specific technique based on

- Only technique known to the analyst
- Technique prescribed by methodology followed by the analyst
- Analysts favorite technique
- Technique selected automatically [28].

A selected technique may not be feasible and applicable within the time frame and the available resources. Therefore, addressing the gap between what is desired and what is available, and selecting the techniques, approaches, and tools in line with the resources can overcome the feasibility and applicability issue [25].

However, appropriate technique selection in distributed projects is very challenging, as stakeholders involved are heterogenous in nature [1, 15]. Tiwari et al. [31] state that considering the influencing parameters of the system such as:

- *Situational characteristics of the project* involving stakeholder types, social environment, domain of the system being developed, scope of the system, analyst ability/skill, approach to be followed, and resource availability
- *Lists of available elicitation techniques*
- *Sources of domain knowledge* such as technical literature, existing implementation, surveys, current and future requirements, and expert advice
- *Available domain knowledge base data* such as existing information, expert knowledge, requirement engineering knowledge base and hypothesis
- A *mapping mechanism* to choose the set of elicitation techniques based on step 1 to 4 can overcome this challenge.

The adoption of standardized data collection tools, which have already been tested and tried in existing projects, also maximizes the data quality. But when the techniques are adapted to the local context, it is essential to conduct the pilot test before using the technique more generally. Doing this will avoid nuisances and awkward situations and reduce biases and errors [25]. The procedures, challenges and best practices associated with "selecting techniques, tools and approaches" are summarized in Table 4.

Table 4. Selecting techniques tools and approaches

How	Challenges	Best practices the practitioners can adopt
Techniques known to the analyst	Feasibility and applicability issues	Ensure the techniques, tools and approaches are in line with the available resources [25]
		Use pilot testing to avoid nuisances [25]
		Consider situational characteristics of the project [31]
Techniques prescribed by the methodology followed by the analyst	May not be best technique	Use the existing information and implementation, surveys, expert advice [31]

3.5 Eliciting the Requirements from the Stakeholders and Other Sources

The requirements for the desired system can be elicited from the identified stakeholders and other sources of requirements using selected tools/techniques and approaches. Jiao and Chen [18] classified the requirements into end-user requirements and functional requirements. The customer/end user requirements tend to be linguistic and usually non-technical. Whereas, the functional requirements tend to have more technical specifications [11].

However, *problem of scope, problem of understanding,* and *problems of volatility* may arise during requirements elicitation [3]. *Problem of scope* refers to not defining the boundaries of the system accurately. When setting the scope, organizations don't always consider the availability of resources, and many organizations don't involve the development team in setting the scope. Hence, the cost and budget requirements are often neglected resulting in over scoping [5]. To avoid the *problem of scope*, the boundary of the new system, its objectives, and the necessary design information must be carefully determined and aligned with the available resources before commencing the interview process.

Problem of understanding refers to poor communication between the requirement analysts and the stakeholders in defining and understanding the requirements. Some projects involve outsourcing and the outsourcing partners can be from non-English-speaking nations. But problems of understanding may arise even if everybody speaks English. Therefore, the necessary information should be expressed in such a way that it improves the communication and understanding between the stakeholders and the analysts. The *problem of understanding* can also be alleviated by institutionalizing prototyping. Demonstrating a prototype to the stakeholders can help bring out the hidden aspects that are initially unknown to the involved stakeholders.

Problem of volatility arises due to changing requirements by the stakeholders. The reason being unforeseen parameters or increased understanding by the stakeholders with the system development. Performing the requirements elicitation tasks iteratively to accommodate changing requirements of the end-users can help overcome the problem of volatility [3, 6].

Another challenge may be reluctant participants. An individual may have unfavorable expectations of the new system, imagining that the new system will make working conditions less enjoyable, burden habitual work processes, or bring about unemployment due to automation of some work processes. Such emotions may negatively affect a stakeholder's decision to share requirements. Therefore, the analyst should act as a psychologist as well as a motivator to predict, detect, and arouse motivating emotions, such as hope and inspiration, in the stakeholders [2].

In inter-organizational projects (IOPs), cross-functional stakeholder requirements are managed across cultural and time zones as well as organizational boundaries. Due to multiple layers of stakeholders, the analyst may have to communicate with end-users through the collocated field support personnel. This indirect communication may lead to increased chances of misinterpretations of stakeholders needs at each level. Therefore, it becomes vital to create stakeholder roles with clear responsibilities, assign these roles to different stakeholders, as well as assign communication responsibilities within each organization involved.

Cultural factors, such as varying attitudes towards hierarchy, and different communication styles also impact the delivery of correct information. Hence, establishment of cultural liaisons can bridge the cultural differences across different sites [9]. The cultural liaisons should work out modalities to enhance trust, like familiarizing themselves with other languages and cultures [29].

Given the sometimes-substantial time differences between the sites, distributed stakeholders continue to rely on asynchronous channels for communication, such as email. But stakeholders may not have a sufficiently good grasp on the language to communicate their viewpoints accurately via email. To alleviate such challenges, the inter-organizational process should be partially synchronized.

Moreover, there may be inconsistencies in the terminology and notations used in the documents across different organizations. To avoid this, the stakeholder groups can define a requirement specification vocabulary and templates for the requirements description at the outset of the project [9]. The procedures, challenges and best practices associated with "eliciting requirements from the selected stakeholders and other sources" is summarized in Table 5.

Table 5. Eliciting requirements from the Selected Stakeholders and other sources

How	Challenges	Best practices the practitioners can adopt
Identified stakeholders	Tacit knowledge	Knowledge triggering [2]
	Unexpected emotions	Supporting psychological and motivational process [2]
	Problem of scoping	Define the boundaries of the desired system [3, 6]
	Problem of understanding	Prototyping [3, 6]
	Problem of volatility	Iterative requirements elicitation process [3, 6]
	Asynchronous communication in IOPs	Partially synchronized process [9, 29]
	Varying attitudes towards hierarchy and communication styles in IOPs	Cultural liaisons [9, 29]
	Lack of direct communication IOPs	Establishment of peer to peer links at project, management, and team levels across distributed sites [9]
Documentation	Inconsistencies in notations and terminology in documentation in IOPs	Requirement specification vocabulary [9]
		Functional design, requirement description templates [9]

4 Discussion and Conclusion

The cumulated tasks, challenges, and best practices in requirements elicitation derived from the literature may help system developers in assessing and mitigating risks and focusing on essential aspects in requirements elicitation. A team consisting of domain expert and domain ignorant analysts can be effective in acquiring the complete and necessary knowledge about the application domain. Solitary requirement exercise and knowledge triggering can overcome the problem of ambiguity. Identifying stakeholders based on project goals objectives, and domain, then filtering the stakeholders based on knowledge and interest, and lastly, prioritizing the identified stakeholders based on the interpersonal skills can help identify the correct stakeholders. Collecting the requirements from different sources increases the credibility of the requirements. When selecting the techniques used for requirement elicitation, it is imperative to do pilot testing and address the gap between what is desired and what is available. Moreover, defining the boundaries of the desired system, institutionalizing prototyping, and psychological and motivational support for the end-users can make the requirements elicitation process more robust. Lastly, establishment of cultural liaisons, peer to peer links at project, management, and team level, and functional design and requirement description templates can overcome the challenges in distributed projects.

Synthesizing from the reviewed studies, we provide detailed explanations of how the fundamental requirements elicitation tasks should be carried out and what the challenges associated with these tasks are. Additionally, we provide recommendations to practitioners for managing the challenges associated with these five fundamental types of requirements elicitation tasks. This study may serve as a starting point for future research on making the requirements elicitation process sufficiently robust to be able to gather the critical system requirements accurately.

Limitations of our study include the fact that we limited our attention to only five most fundamental types of requirements elicitation tasks suggested by Zowghi and Coulin [35], while there are many other requirements elicitation tasks listed in the literature. Thus, future research may look at these other tasks, such as for example testing the correct set of requirements by prototyping [30]. There may be also be specific tasks that don't apply to specific projects. Future research may look at additional research question, such as which task/activities apply to which type of projects.

References

1. Alexander, I.: Does requirements elicitation apply to open source development? In: OSSG (2009)
2. Apshvalka, D., Donina, D., Kirikova, M.: Understanding the problems of requirements elicitation process: a human perspective. In: Wojtkowski, W., Wojtkowski, G., Lang, M., Conboy, K., Barry, C. (eds.) Information Systems Development, pp. 211–223. Springer, Boston (2009). https://doi.org/10.1007/978-0-387-68772-8_17
3. Ashraf, I., Ahsan, A.: Investigation and discovery of core issues concerning requirements elicitation in Information Technology industry and corresponding remedial actions (an inductive case study of Pakistan's IT industry). In: 2010 IEEE 17th International Conference on Industrial Engineering and Engineering Management (IE&EM), pp. 349–353). IEEE, October 2010
4. Ballejos, L.C., Montagna, J.M.: Method for stakeholder identification in inter-organizational environments. Requir. Eng. **13**(4), 281–297 (2008)
5. Bjarnason, E., Wnuk, K., Regnell, B.: Are you biting off more than you can chew? A case study on causes and effects of overscoping in large-scale software engineering. Inf. Softw. Technol. **54**(10), 1107–1124 (2012)
6. Bormane, L., Gržibovska, J., Bērziša, S., Grabis, J.: Impact of requirements elicitation processes on success of information system development projects. Inf. Technol. Manag. Sci. **19**(1), 57–64 (2016)
7. Chakraborty, S., Sarker, S., Sarker, S.: An exploration into the process of requirements elicitation: a grounded approach. J. Assoc. Inf. Syst. **11**(4), 212 (2010)
8. Cheng, B.H., Atlee, J.M.: Research directions in requirements engineering. In: 2007 Future of Software Engineering, pp. 285–303. IEEE Computer Society, May 2007
9. Damian, D.: Stakeholders in global requirements engineering: lessons learned from practice. IEEE Softw. **24**(2), 21–27 (2007)
10. Davis, C.J., Fuller, R.M., Tremblay, M.C., Berndt, D.J.: Communication challenges in requirements elicitation and the use of the repertory grid technique. J. Comput. Inf. Syst. **46**(5), 78–86 (2006)

11. Dias, R., Cabral, A.S., López, B., Belderrain, M.C.N.: The use of cognitive maps for requirements elicitation in product development. J. Aerosp. Technol. Manag. **8**(2), 178–192 (2016)
12. Ferrari, A., Spoletini, P., Gnesi, S.: Ambiguity and tacit knowledge in requirements elicitation interviews. Requir. Eng. **21**(3), 333–355 (2016)
13. Friedrich, W.R., Van Der Poll, J.A.: Towards a methodology to elicit tacit domain knowledge from users. Interdisc. J. Inf. Knowl. Manag. **2**(1), 179–193 (2007)
14. Fuentes-Fernández, R., Gómez-Sanz, J.J., Pavón, J.: Understanding the human context in requirements elicitation. Requir. Eng. **15**(3), 267–283 (2010)
15. Gill, K.D., Raza, A., Zaidi, A.M., Kiani, M.M.: Semi-automation for ambiguity resolution in Open Source Software requirements. In: 2014 IEEE 27th Canadian Conference on Electrical and Computer Engineering (CCECE), pp. 1–6. IEEE, May 2014
16. Glinz, M., Wieringa, R.J.: Guest editors' introduction: stakeholders in requirements engineering. IEEE Softw. **24**(2), 18–20 (2007)
17. Hadar, I., Soffer, P., Kenzi, K.: The role of domain knowledge in requirements elicitation via interviews: an exploratory study. Requir. Eng. **19**(2), 143–159 (2014). 2014 International Conference on Computing for Sustainable Global Development (INDIACom), pp. 151–155. IEEE
18. Jiao, J., Chen, C.H.: Customer requirement management in product development: a review of research issues. Concurr. Eng. **14**(3), 173–185 (2006)
19. Majumdar, S.I., Rahman, M.S., Rahman, M.M.: Thorny issues of stakeholder identification and prioritization in requirement engineering process. IOSR J. Comput. Eng. (IOSR-JCE) **15**(5), 73–78 (2013)
20. Morales-Ramirez, I.: On exploiting end-user feedback in requirements engineering. In: 19th International Working Conference on Requirements Engineering: Foundations for Software Quality, Doctoral Symposium Programme, pp. 223–230 (2013)
21. Niknafs, A., Berry, D.M.: An industrial case study of the impact of domain ignorance on the effectiveness of requirements idea generation during requirements elicitation. In: 2013 21st IEEE International Requirements Engineering Conference (RE), pp. 279–283. IEEE, July 2013
22. Pa, N.C., Zin, A.M.: Requirement elicitation: identifying the communication challenges between developer and customer. Int. J. New Comput. Archit. Appl. (IJNCAA) **1**(2), 371–383 (2011)
23. Pacheco, C., Tovar, E.: Stakeholder identification as an issue in the improvement of software requirements quality. In: Krogstie, J., Opdahl, A., Sindre, G. (eds.) CAiSE 2007. LNCS, vol. 4495, pp. 370–380. Springer, Heidelberg (2007). https://doi.org/10.1007/978-3-540-72988-4_26
24. Pandey, D., Suman, U., Ramani, A.K.: An effective requirement engineering process model for software development and requirements management. In: 2010 International Conference on Advances in Recent Technologies in Communication and Computing (ARTCOM), pp. 287–291. IEEE, October 2010
25. Peersman, G.: Overview: data collection and analysis methods in impact evaluation. UNICEF Office of Research-Innocenti (2014)
26. Qureshi, N.A., Seyff, N., Perini, A.: Satisfying user needs at the right time and in the right place: a research preview. In: Berry, D., Franch, X. (eds.) REFSQ 2011. LNCS, vol. 6606, pp. 94–99. Springer, Heidelberg (2011). https://doi.org/10.1007/978-3-642-19858-8_11
27. Razali, R., Anwar, F.: Selecting the right stakeholders for requirements elicitation: a systematic approach. J. Theor. Appl. Inf. Technol. **33**(2), 250–257 (2011)
28. Sachdeva, S., Malhotra, M.: Requirement elicitation of large web projects. Int. J. Eng. Comput. Sci. **3**(07), 6880–6887 (2014)

29. Shah, Y.H., Raza, M., UlHaq, S.: Communication issues in GSD. Int. J. Adv. Sci. Technol. **40**(2012), 69–76 (2012)
30. Sharma, S., Pandey, S.K.: Requirements elicitation: issues and challenges, March 2014
31. Tiwari, S., Rathore, S.S., Gupta, A.: Selecting requirement elicitation techniques for software projects. In: 2012 CSI Sixth International Conference on Software Engineering (CONSEG), pp. 1–10. IEEE, September 2012
32. Wellsandt, S., Hribernik, K.A., Thoben, K.D.: Qualitative comparison of requirements elicitation techniques that are used to collect feedback information about product use. Procedia CIRP **21**, 212–217 (2014)
33. Zhang, Z.: Effective requirements development-a comparison of requirements elicitation techniques. In: Berki, E., Nummenmaa, J., Sunley, I., Ross, M., Staples, G. (eds.) Software Quality Management XV: Software Quality in the Knowledge Society, pp. 225–240. British Computer Society (2007)
34. Zhao, M., Ji, Y.: Challenges of introducing participant observation to community health research. ISRN Nurs. **2014**, 7 (2014)
35. Zowghi, D., Coulin, C.: Requirements elicitation: a survey of techniques, approaches, and tools. In: Aurum, A., Wohlin, C. (eds.) Engineering and Managing Software Requirements, pp. 19–46. Springer, Heidelberg (2005). https://doi.org/10.1007/3-540-28244-0_2

Effective Decision-Making in Supply Chain Management

Henryk Krawczyk[1(✉)] and Narek Parsamyan[2(✉)]

[1] Faculty of Electronics,Telecommunications and Informatics,
Gdansk University of Technology, Gabriela Narutowicza 11/12, Gdansk, Poland
hkrawk@pg.edu.pl
[2] Centre of Informatics - Tricity Academic Supercomputer and Network CI TASK,
Gdansk University of Technology, Gabriela Narutowicza 11/12, Gdansk, Poland
narparsa@pg.edu.pl

Abstract. Proper decision-making in Supply Chain Management (SCM) is crucial for an appropriately functioning mechanisms. The paper presents how IT technologies can impact on an organization and process realization. Especially Service Oriented Architecture (SOA) standard and the Cloud Computing (CC) paradigms are taken into account. A general model of decision-making is proposed and based on a specific practical example is analyzed using Business Process Model and Notation (BPMN). It was shown that among the available product ordering propositions we can choose that which minimizes total product waiting time. In consequences, we can reduce wasted time which means greater efficiency in decision-making processes.

Keywords: Supply Chain Management · Decision-making process
Business process modeling and improvement

1 Introduction of the General Model of SCM

A supply chain is the network of suppliers, distributors, and subcontractors used by a manufacturer to organize work-in-progress and offer final products to distributor and intermediates [1]. The last ones are responsible for delivering the final products to the market, where sellers offer them to their clients. Such a general model is presented in Fig. 1.

To coordinate all activities or enterprises in such a complex model a suitable collaborative framework must be realized. It is based on so-called Supply Chain Management (SCM) [10]. SCM is a set of approaches that efficiently integrates suppliers, manufacturers, warehouses and stores for planning, implementing, and controlling of the materials and information flows from origin to the point of destination; so that merchandise is produced and distributed in the right quantities, to the right location, and at the right time, in order to minimize system-wide costs, and to provide providing satisfying service level requirements. There are

© Springer Nature Switzerland AG 2018
S. Wrycza and J. Maślankowski (Eds.): SIGSAND/PLAIS 2018, LNBIP 333, pp. 45–53, 2018.
https://doi.org/10.1007/978-3-030-00060-8_4

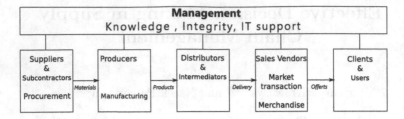

Fig. 1. The general model of a supply chain

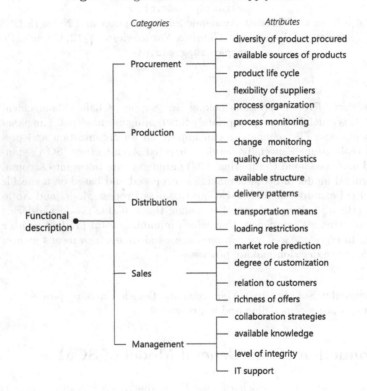

Fig. 2. The basic attributes of functional categories of a supply chain

a variety of supply chain models and their implementations, this means also an existing variety of solutions of supply chain management [8].

In Fig. 2 is presented the main functions of a supply chain: procurement, production, distribution, sales and management, and attributes corresponding to each of them. At the earliest stage of a supply chain, the information flow among its organizational structures and its members (actors) were paper-based. It caused that information was often overlooked, and its value was rarely/clearly understood. IT infrastructure capability just provides new possibilities related to effective implementation of many cross-functional data flows and processes. It causes that data, information and knowledge, become a crucial factor in managers' abilities, and the supply chain power was shifted from manufacturers to retailers. Many companies share information from many retail outlets with manufacturers or other major suppliers. In consequence, supply chains have new distinct advantages, such as cost reduction, higher productivities, improvement in market selling strategies. Many companies use ERP systems [1] as the core of their IT infrastructures. It allows them to achieve a higher level of integration by utilizing shared data, better understanding rules for on-line data access and high discover of customer's needs. In the paper, we limit our consideration only to decision-making, and consider how Service Oriented Architecture (SOA) can improve supply chain functionality.

Depending on the category of supply chain we can distinguish corresponding attributes to them (Fig. 2). The presented attributes are basically universal, it means they can refer to more than one group of supply chain categories, and can be used as a decision-making factor. In the next section, we describe decision-making models and their suitability for supply chain management. Further, we consider the specific examples of decision-making in delivery patterns to achieve market transitions into the Just-in-time (JIT) approach [7].

2   IT Services Supporting Decision Making SCM

Decision-making is one of the essential aspects of Supply Chain Management (SCM). It relates to the realized processes when we met a problem and tried to solve it. In Fig. 3 we present a possible scenario of decision-making in such situations. The proposed scenario consists of four of the following steps:

- recognize and identify the problem,
- gather information from different sources about the problem and its possible solutions,
- generate possible solutions for the problem,
- choose the adequate solution in a considered context describing the current situation.

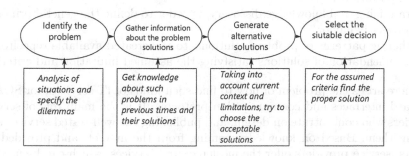

Fig. 3. The supply chain scenario of decision-making

There are two kinds of problems met in SCM. The first is SCM problem-oriented, that corresponds to activity directly related to aspects shown in Fig. 3. The second one, called human-oriented problems referred to personal conflicts, unproductive communication's, or low member motivations. In both situations, the proper solutions should be found, and a suitable decision should be made, which will eliminate or reduce negative influences on an effective management process. It depends very often on available knowledge, efficient communication, and human creativity, and ethics. In complex situations, some experts can be included in the decision-making process [3].

To compare Figs. 1 and 3 we can note, that the decision making scenario corresponds to the supply chain concept on different levels of description, where their basic activities are quite different. It seems that we consider nested SCM's, however, it is not easy to achieve a high level of such integration. One serious possibility is to create and enforce an exclusive set of IT services supporting some activities of the entire SCM. Two IT technologies create such possibilities: Service Oriented Architecture (SOA) and Cloud Computing (CC). The concept of SOA has emerged as a powerful architectural paradigm referring to the needs of companies in competitive environments. Services package application functionality and make it available through interfaces; due to that the hide implementation detail, and separate service description from the execution environment. The Extensible Markup Language (XML) standards are used for their description. They may be a functional part of some other services, and in that sense they are composable. The rise of an abstraction level for service description allows the narrowing of the gap between its implementation and its actual business function. In consequence, a business-purpose oriented combination of several services become a complex service composition or an orchestration representing an implemented business process [9]. An example set of IT services supporting a decision-making process combines the following group of activities (see Fig. 3):

- monitoring of chosen human beings,
- environment or platform actions,
- store implementation about how-to in SCM, and also about met situations or expectations,
- data mining and knowledge discovery, in order to define the general principle of proper behavior,
- gathering patterns of problem solutions, to the created available repository,
- recommendation of solutions satisfying the assumed limitation and criteria.

There arises the problem of how to find such kinds of IT services, for SOA's, standard interfaces and communication protocols. It is the main role of service providers who concentrate on design, and publishing the well-tested services and hosting them. Based on knowledge coming from the market, and provided by brokers, service providers offer the most required services, which can be used in different computing environments, and can be chosen directly by the users to create the required business scenarios: such business scenarios represent flow-charts of different human activities supported completely or partially by the available set of IT services. The well-known software engineering oriented on design and implementation of IT services and applications is named Service Oriented Software Engineering (SOSE). The SOSE approach requires solutions connected with effective virtualization and interoperability, which are provided by Computing Clouds (CC) [4]. Three layers of CC: Infrastructure as a Service (IaaS), Platform as a Service (PaaS), and Software as a Service (SaaS) create multitenant environments for services and applications execution, design, and implementation, and also support service security and availability. User's business relies on such third-party service providers where attributes of service quality play an important role in the market. Therefore service and cloud users should check the quality before using particular services. For such attributes in general, the exact physical location of data and services are unknown. It may be local, domestic or foreign. In such cases protecting an audit is required to eliminate some disasters or cyber-attacks. Service providers establish SLA's which need be checked by users, to ensure the proposed conditions are satisfied.

3   Examples

Figure 2 describes attributes related to the basic categories of SCM. There are many publications [2] where each of the attributes were analyzed separately, and their impact on SCM functionality was shown. Utilization of new IT technology (SOA and CC) creates new opportunities to make analysis more horizontal where different attributes are simultaneously considered. Such an approach is much more required for management (Fig. 3) where the suitable knowledge should be collected, the proper collaboration strategy should be organized, and minimal integrity of all components of SCM should be assured. In consequence, the same set of IT services should be prepared to support such purposes of SCM. Below, to illustrate such an approach we consider only one problem how to realize inventory management strategy in order to achieve continuous accessibility of products by clients in a market. Such a strategy should be characterized by minimal cost, proportional to the total time τ of all product waiting times, i.e.

$$\tau = \sum_{i=1}^{N} * \sum_{j=1}^{M_i} t_{ij} \qquad (1)$$

where:

- N - is the number of different categories of products waiting in the considered store
- M_i- is the number of products waiting in the store
- t_{ij} - is the waiting time of j-th product belonging to i-th category.

The number N, M_i, and values t_{ij} are changing over time, because some products are selling and the new ones are delivered from the warehouse. Supply chain problem is how to organize product delivery to the store to achieve that the τ value be as low as possible. In other words we should choose the strategy which satisfies the below criterion:

$$0 < \tau \leq const \tag{2}$$

Criterion (2) means that in each moment of time t, total time τ does not exit the assumed value const. We may take into consideration also the mean total time calculated for all products, or each category of products.

The development of information technologies brings great opportunities for modern companies which want to develop digitally. A good example of positive use of computer science in enterprises business activities is computer support in the supply management and decision-making process. We will use BPMN 2.0 for the purposes of analyzing the processes taking place in the company dealing with the product sales. The business models created in this notation will reflect the actual flow of information in the decision process, from identifying the problem to making the decision (Fig. 3). The BPMN standards characterize a process workflow with a set of graphical tools, in order to provide the most accurate description of the process, so that it would be universal, transparent and, most importantly, understandable. The OMG (Object Management Group) in 2011 published Standard BPMN 2.0 and it is currently a widely-used and well-known worldwide standard for the graphic representation of business models [5].

The model in Figs. 4 and 5 presents a company that has a supply chain management issue, in particular, an inventory management problem. We will consider how to optimize the inventory management process with IT support, in particular through using services, and the SOA approach. On the first model (Fig. 4) we consider the scenario where the activities are performed manually by the people in the company, they have only access to digital data, but the operations are carried out manually. In contrast, in the second model (Fig. 5) some activities are performed automatically and using a cloud computing environment.

In Fig. 4 is presented the classical inventory management model in a retail company. In this model, we assumed that the key decision is to determine the number of ordered goods and the time of their delivery to the warehouse. The process starts when the manager needs to check the daily inventory level. The manager is requesting preparation of the order proposal for products. Then the analysis process starts. Firstly, analysts gather the information' from the database, where is included precise information about supply chain functionality e.g.:

- product life cycle,
- flow of goods (products came in, and came out, cycle),
- process organization,
- inventory level,
- delivery patterns,
- historical statistics.

The analysts get information (e.g. in .xls format) from a database and in the next step they analyse the data in order to maintain the optimum inventory level in the warehouse. Due to companie's desire to optimize costs, it is

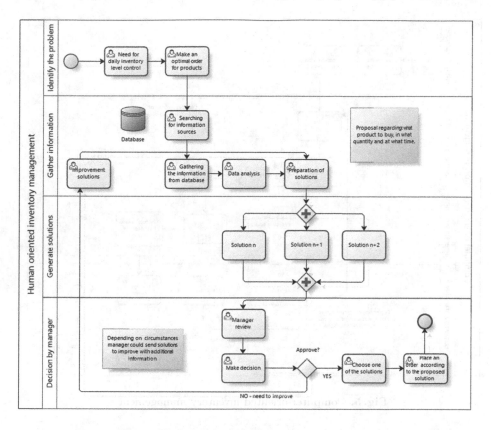

Fig. 4. Human-oriented inventory management

important there shouldn't be too many products in the warehouse (because of increasing cost of maintenance) and/or too little products in the warehouse (lack of goods may cause the customer to resign from the order). Based on historical data and using predictive models (e.g. linear models, decision-tree) the analysts prepare solutions, including the recommendation for goods order for the next days. Then the manager is reviewing the report and based on his/her knowledge and experience can accept, reject or request for solutions repeated (with additional remarks). The cycle is completed when the manager considers that the proposed solutions are sufficient to take optimal decisions. The model ends when the manager placed the order for goods.

The key difference between the models presented in Figs. 4 and 5 is the system IT instead of human activities. In the Fig. 5, the manager is using the computer application in the form of IT services for preparing the order for products. It is worth noting that this type of IT services can be performed using services outsourcing available in the computing cloud platform. The services, in general, replace the human work and automatically analyze the sources and then the knowledge base. The analysis process taking place in Fig. 5 is almost the same

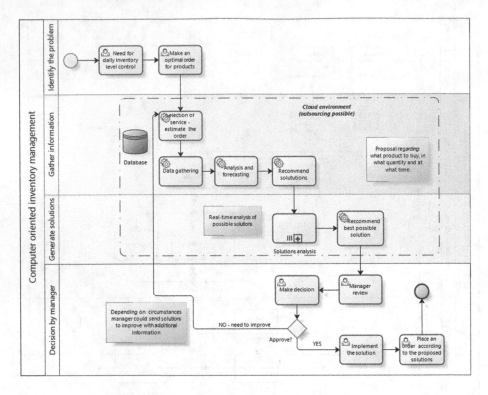

Fig. 5. Computer-oriented inventory management

as in the model in Fig. 4; with this difference that calculations are performed automatically in real-time by using Neutral Neural Networks (NNN) [6]. Thanks to the possibility of analyzing hundreds of solutions and parallely comparing them with each other, the computer recommends the most beneficial, and the most effective, solution. The decision about what to do still remains in the hands of the manager, however, thanks to the deployment of IT services, the risk of choosing an unfavorable solution is minimized.

4   Final Remarks

Analyzing processes of decision-making in supply chain management we need big knowledge to describe them in an understanding way. Producer, distributors, sellers, and clients have different parts of such knowledge, but it is not generally available. Due to technology we can register some data referring to such processes and then using data mining we can discover some rules and trends. However diversity of supply chain management causes that understanding the processes of decision-making is still very distant. Additionally, as it was proposed in the paper, use of the BPMN standard can facilitate the process, and gives the possibility for a deeper analysis of the warehouse management process.

The paper highlights the essence of the use of service-oriented systems in supply chains and decision-making processes. It should be noted that the use of IT systems not only shortens the duration of tasks and calculations execution, but also ensures higher accuracy and efficiency. In addition, process support through the outsourcing of cloud computing services enables a rapid development of modern recommendation services e.g. forecasting based on data mining, without the need to invest large financial and technical resources.

Large use of SOA and cloud computing additionally with Internet of Things and industry 4.0/5.0 technology will shape and support SCM. Combining different modern digitalization processes we transform traditional supply chains to more flexible ones. Moreover, artificial intelligence integrate with business intelligence will add to them learning processes, which in consequences will create a new supply chains category called: smart supply chains [12].

References

1. Botta-Genoulaz, V., Millet, P.A., Grabot, B.: A survey on the recent research literature on ERP systems. Comput. Ind. **56**, 510–522 (2005)
2. Erbes, J., Graupner, S., Motahari-Nezhad, H.R.: The future of enterprise IT in the cloud. IEEE Internet Comput. **45**, 66–72 (2012)
3. Krawczyk-Bryka B., Krawczyk H.: The preferable way of decision making in IT teams. AGH J.-Decis. Mak. Manuf. Serv. (in review process)
4. Krawczyk, H., Nykiel, M., Proficz, J.: Mobile offloading framework: solution for optimizing mobile applications using cloud computing. In: Gaj, P., Kwiecień, A., Stera, P. (eds.) CN 2015. CCIS, vol. 522, pp. 293–305. Springer, Cham (2015). https://doi.org/10.1007/978-3-319-19419-6_28
5. Krawczyk, H., Parsamyan, N.: Enterprise activities modeling by BPMN notation. Inf. Syst. Manag. **6**(2017), 203–212 (2017)
6. Linton, J.D., Klassen, R., Jayaraman, V.: Sustainable supply chains: an introduction. J. Oper. Manag. **25**(6), 1075–1082 (2007)
7. Qanbari, S., Li, F., Dustdar, S.: Toward portable cloud manufacturing services. IEEE Internet Comput. **6**(11/12), 77–80 (2014)
8. Simchi-Levi, D., Simchi-Levi, E., Kaminsky, P.: Designing and Managing the Supply Chain: Concepts, Strategies, and Cases. McGraw-Hill, New York (1999)
9. Wrycza, S. (ed.): Informatics Economy. PWE (2010)
10. Rozga, A., eriN, Ante, L.: Business Intelligence and Supply Chain Management. In: International Conference on Information technology (ICIT) (2013)
11. Liu, E., Yang, S., Wang, S., Bai, E.: A scheduling model of logistics service supply chain based on the time windows of the FLSPs operation and customer requirement. Annu. Oper. Res. **257**, 183–206 (2017)
12. Wu, L., Yue, X., Jin, A., Yen, D.C.: Smart supply chain management: a review and implications for future research. Int. J. Logist. Manag. **27**, 395–417 (2016)

Systems Acceptance and Usability

The Acceptance of Mobile Technology in Knowledge Providing Model from the Perspective of User's Characteristics

Janusz Stal[✉] and Grażyna Paliwoda-Pękosz

Department of Computer Science, Cracow University of Economics,
Rakowicka 27, 31-510 Kraków, Poland
{janusz.stal, paliwodg}@uek.krakow.pl

Abstract. The study presents the results of an investigation into the acceptance of mobile technology usage for knowledge providing in different contexts: (1) educational/work environment, (2) mobile users' professional background (related and non-related to information and communication technology (ICT)), and (3) individuals' age. The Technology Acceptance Model (TAM) was adjusted for the purpose of explaining and predicting mobile users' intentions. Then, the model was empirically examined using the structural equation modeling (SEM) with the dataset of 303 individuals. The results reveal differences in mobile technology usage acceptance in different subgroups of respondents. However, similar results of hypothesis testing were obtained for respondents without ICT background and older respondents. The research findings imply the necessity of adjusting the content distributed via mobile devices to recipients' age, their ICT skills, and context of usage, i.e. education versus work environment.

Keywords: Technology Acceptance Model (TAM) ·
Structural equation modeling (SEM) · Mobile technology ·
Knowledge providing

1 Introduction

The knowledge-based economy (KBE) is a concept that emerged in the 1980s, defining the stage of economic development where knowledge plays a decisive role in stimulating social and economic development, which determines the level and pace of changes [9, 18, 19, 22]. KBE is the dominant post-industrial economic development paradigm which defines an economy based on production, distribution, and the use of knowledge and information, which are the most important factors leading to economic growth [13]. It is based on advanced technologies and industries employing qualified employees. Such phenomena arouse a keen interest of international organizations, namely: OECD, European Union and World Bank [7, 19]. The latter emphasizes the importance of knowledge which should be a key factor in a country's development strategy and is based on [32]: (1) education and training, (2) information and telecommunications infrastructure, (3) innovation, and (4) the overall business and governance framework. It seems that one of the key elements contributing to the

© Springer Nature Switzerland AG 2018
S. Wrycza and J. Maślankowski (Eds.): SIGSAND/PLAIS 2018, LNBIP 333, pp. 57–67, 2018.
https://doi.org/10.1007/978-3-030-00060-8_5

development of a knowledge-based economy is the development of technologies that process, store and transmit data in electronic form, which enables knowledge delivery.

Mobile technology is one of those whose rapid development can be observed especially in the last decade. The works on mobile operating systems used today, completed in 2006/2007 [10], resulted in numerous benefits and possibilities of utilizing mobile devices (research carried out in this article places the main emphasis on the use of smartphones) in various domains providing access to information, processes, and communication anytime and anywhere [3, 17, 20, 23]. With a mobile connection, employees were given an opportunity to access corporate resources anytime and anywhere, which led to increasing their efficiencies owning to enhanced communication and connectivity [30]. Considering its properties (particularly access to information anytime and anywhere), mobile technology should operate well in the process of gathering, provision, and the use of knowledge [28, 29].

It should be, however, taken into account that several factors may affect an individuals' attitude to the use of mobile technology. Firstly, the context of using mobile devices might be important, i.e. education versus work environment. Hence, it might be interesting to investigate students' perception of mobile technology usage and employees' attitude towards this technology. Secondly, the users' familiarity with modern technologies, in particular information and communication technology (ICT), might be an important factor in mobile technology utilization for knowledge providing. Finally, the users' age might influence their attitude towards mobile technology usage. This is exemplified by prior research showing that age might be an important factor, having an impact on people's attitudes towards ICT. In particular, older generations might perceive different risks related to ICT use than younger people [27]. Furthermore, older users might experience greater difficulties with business software adoption and use compared with their younger counterparts [26]. Age may also play an important role in employees' perception of threats to enterprise system adoption and solutions to the problems encountered during this process [25].

In the context of the mentioned issues, this paper seeks to address the question whether mobile devices can be an effective tool for providing knowledge in the educational or professional zone and what factors would affect this technology's acceptance. It is worth mentioning that a considerable number of existing studies deal primarily with the use of mobile technology for providing knowledge in education [1, 6, 14]. To the best of the authors' knowledge, the case of using mobile technology for providing and acquiring knowledge taking into account aforementioned factors (age, ICT familiarity, educational/professional experience) has not been given considerable attention by researchers in the past, and this motivated the present study. This paper is organized as follows. The next section provides an overview of the Technology Acceptance Model (TAM), develops a research model and hypotheses. Next, research results and analysis are presented. The paper concludes with a summary and future research directions.

2 Research Model and Hypotheses Formulation

In the domain of information systems theory, the Technology Acceptance Model (TAM), derived from the theory of reasoned action [8], gained considerable support in understanding the process of new technology acceptance [4]. Its purpose is to predict the acceptability of a technology to be used and to identify the modifications which must be brought to the system in order to make it acceptable to users. The original TAM consists of the following determinants for system use: PEOU (perceived ease of use), PU (perceived usefulness), ATU (attitude towards using), BI (behavioral intention to use), and AU (actual system use). According to an investigation by Davis et al. [5], the explanatory power of TAM is equally good without using ATU construct. As later studies show, it has become ubiquitous to exclude the attitude construct from the original TAM. Simultaneously, some researchers emphasize the need of using additional variables in TAM to complete the understanding of the phenomenon studied [33]. Prior studies show that researchers have proposed more than 70 external variables for PU and PEOU to reveal how the individuals' perceptions are formed. In their study, Yousafzai et al. [34] have grouped all existing external variables into four categories: (1) organizational characteristics, (2) system characteristics, (3) user personal characteristics, and (4) other variables. In vein of these studies, we found a need of exploring more deeply the system characteristic as the most relevant to our research.

Numerous studies show considerable benefits of using mobile technology in various domains [12, 21]. However, a number of limitations should be taken into account [16, 31], namely: (1) the cost of mobile internet access, (2) network speed and reliability, (3) access to mobile internet, (4) content optimization for correct display on mobile devices, (5) matching the content to the expectations of the mobile user, and (6) physical attributes of mobile devices (small mobile device display dimensions, lack of physical keyboard/mouse). These factors, in our opinion, can have a significant impact on the perceived ease of use and usefulness of mobile technology. Taking into account aforementioned factors and following the Yousafzai et al. [34], a set of external variables have been included in the proposed research model (Fig. 1). The external variable Access to Information (AI) refers to the possibility of access to mobile information resources (mobile network coverage, cost and speed) and has an impact on the perceived usefulness of mobile technology in knowledge providing. Information Quality (IQ) highlights the importance of content optimization intended for mobile devices (content tailored to the expectations of a mobile user) and affects PU and PEOU constructs. The last of external variables, Information Navigation (IN), refers to the possibility of navigation in the content presented on a mobile device and determines the ease of use of a mobile device (PEOU). In our opinion, the relationships between AI and PEOU, as well as between IN and PU are not justified. Presentation/displaying of information depends to a large extent on physical activity. Hence, the ease of use is a key factor here and its perception influences the use of mobile devices in providing knowledge. On the other hand, the usefulness of information depends on its content, hence the perceived usability of mobile devices should depend on access to information. Based on the proposed research model and taking into consideration previous studies, we defined a set of hypotheses describing the relationships between the constructs of the proposed model (Table 1).

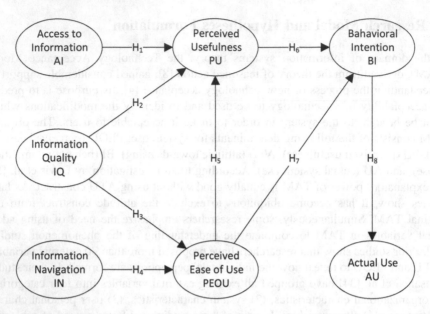

Fig. 1. The proposed research model of mobile technology acceptance for knowledge providing.

Table 1. Research hypotheses.

Hypothesis	Path	Statement
H₁	AC → PU	Access to mobile information has a significantly positive impact on the perceived usability of mobile devices for providing information
H₂	IQ → PU	The quality of mobile information has a significantly positive impact on the perceived usability of mobile devices for providing information
H₃	IQ → PEOU	The quality of mobile information has a significantly positive impact on the perceived ease of use of mobile devices for providing information
H₄	NV → PEOU	The characteristics of navigation of mobile information has a significantly positive impact on the perceived ease of use of mobile devices for providing information
H₅	PEOU → PU	The perceived ease of use of mobile devices has a significantly positive effect on their perceived usability for providing information
H₆	PU → BI	The perceived usefulness of mobile devices for providing information has a significantly positive impact on the intent of their use
H₇	PEOU → BI	The perceived ease of use of mobile devices for providing information has a significantly positive impact on the intent of their use
H₈	BI → AU	The behavioral intent to the use of mobile devices for providing information has a significantly positive impact on actual mobile device use

3 Research Methodology

The current research is an extension of our previous study (forthcoming) in which we developed and tested the TAM model for the whole population of respondents. For the purpose of comprehensibility, we present here briefly the main steps of our research approach. In order to verify the research hypothesis using structural equation modeling (SEM), we first developed a measure model that was reflected in 21 questions of the elaborated questionnaire (each of the seven construct in our research model was represented by three items in the questionnaire that were measured on a seven-point Likert-type scale ranging from 1 – "strongly disagree" to 7 – "strongly agree" (see Table 2). Additionally, we asked the respondents to provide information vital for our current study, i.e. age, types of activities conducted (related or non-related to IT), and the type of environment in which they operate (education or work). For the purpose of obtaining the general characteristics of respondents, we included in the questionnaire items concerning the respondents' gender, frequency of using a smartphone and a mobile Internet connection, as well as the type of education.

Table 2. Measurement items in the survey questionnaire.

Constructs (latent variables)	Measurement items	Questions
Access to Information AI	AI1	In my opinion, the speed of mobile Internet is satisfactory
	AI2	In my opinion, the availability of mobile Internet is satisfactory
	AI3	In my opinion, the access cost to the mobile Internet is satisfactory
Information Quality IQ	IQ1	I think that content intended for smartphones is optimized for their correct display
	IQ2	I think that content displayed on smartphones is tailored to the expectations of a mobile user
	IQ3	I think that it is possible to effectively present any type of content on a smartphone
Perceived Ease of Use PEOU	PEOU1	I think that it's easy to use a smartphone
	PEOU2	I think it would be easy for me to use a smartphone to get information
	PEOU3	In general, I think that using a smartphone to get information would be easy
Information Navigation IN	IN1	I think that small dimensions of a smartphone do not constitute an obstacle to effective navigation of the displayed information
	IN2	I think that the lack of a traditional keyboard or mouse is not an obstacle to effective navigation of the presented information

(continued)

Table 2. (*continued*)

Constructs (latent variables)	Measurement items	Questions
	IN3	In general, I think that the technical parameters of a smartphone should not be an obstacle to getting acquainted with the presented content
Perceived Usefulness PU	PU1	I believe that using a smartphone to obtain information can speed up the implementation of tasks
	PU2	I believe that using a smartphone to obtain information can improve my work efficiency
	PU3	In general, I think that using a smartphone to get information can be useful in my work
Behavioral Intention BI	BI1	I intend to use a smartphone to retrieve information in the future
	BI2	I intend to use a smartphone to get information as often as possible
	BI3	I intend to use a smartphone to obtain information to support my work
Actual Use AU	AU1	I used a smartphone to get information during the last week
	AU2	I used a smartphone to get information during the last month
	AU3	In general, I use a smartphone to get information

The questionnaire was implemented in the cloud using the Google Forms service and made publicly available between March and May 2018. The link to this questionnaire was published on the university e-learning platform, social media sites, and was distributed to other potential respondents in emails.

We followed the two-step approach for Structural Equation Modelling (SEM) proposed by Anderson and Gerbing [2]. The analysis was performed using the Statistica software package (http://www.statsoft.com/Products/STATISTICA-Features). Firstly, a Confirmatory Factor Analysis (CFA) was performed that resulted in the development of the measurement model. Secondly, the proposed structural model that includes causal relationships between constructs (the model hypothesis) was tested using SEM.

4   Results

4.1   Data

In total, the questionnaire was completed by 303 respondents. The characteristics of the respondents are presented in Table 3. A slightly greater share in the respondents' population are male than female. More than half of the respondents are below 30 years old; more than 60% of respondents declared that their activities (work, studies) are related to ICT.

Table 3. Respondents' characteristics.

Variable	No.	%
Gender		
Female	126	42%
Male	177	58%
Age		
<20	10	3%
21–30	185	61%
31–40	30	10%
41–50	37	12%
>50	41	14%
Frequency of use		
1–2 times a day	22	7%
1–2 times a week	10	3%
numerous times a day	241	80%
Constantly	27	9%
Never	3	1%
Professional experience		
Student	57	19%
Employee	96	32%
Student and employee	133	44%
Other	17	5%
Education		
Humanities	21	7%
Economics	51	17%
Sciences	18	6%
Sciences related to ICT	173	57%
Other	40	13%
Activities/work		
Related to ICT	187	62%
Non related to ICT	116	38%

4.2   Reliability of the Measurement Model

The Cronbach's alpha for the seven latent variables ranged from 0.57 to 0.96, being the lowest for Information Navigation (0.57), Information Quality (0.61) (considered to be poor but acceptable in the early stage of the measurement model development [24]), and the highest for Actual Use (0.96). The correlation between latent variables ranged from 0.145 to 0.916 and, except one (IN-AI), was significant at p = 0.01. The fit indices of the measurement model are close to the recommended level [15] (χ^2 = 563.293, significant at p = 0.000, but this result is acceptable due to its sensitivity to the sample size [11]; χ^2/degree of freedom = 3.3; SRMR (Standardized Root Mean Residual) = 0.073; CFI (Comparative Fit Index) = 0.893; NFI (Normed Fit Index) = 0.856).

4.3   Hypothesis Testing

The results of the structural model testing in the different subgroups of respondents are presented in Table 4. In the groups of younger respondents and students all hypotheses were accepted whereas in the group of older respondents, hypothesis H2 and H4 were rejected. Similarly, in the group of respondents whose activities are not related to ICT, hypothesis H2 and H4 were rejected, whereas hypothesis H1 and H2 were rejected in the group of respondents whose activities are related to ICT and who are employed.

Table 4. The results of hypothesis testing in different subgroups of respondents.

	No	H_1	H_2	H_3	H_4	H_5	H_6	H_7	H_8
		AI → PU	IQ → PU	IQ → PEOU	IN → PEOU	PEOU → PU	PU → BI	PEOU → BI	BI → AU
All	303	−0.026*	0.067*	0.407	0.248	0.355	0.991	0.458	0.848
Age									
<= 30	195	0.152	0.143	0.295	0.217	0.298	0.896	0.538	0.850
>30	108	−0.142	0.052*	0.580	0.119*	0.294	1.161	0.436	0.865
Work									
non related to ICT	116	−0.081	0.088*	0.605	0.077*	0.153	1.448	0.445	0.842
related to ICT	187	0.046*	0.071*	0.311	0.239	0.519	0.817	0.549	0.845
Status									
students	190	0.113	0.127	0.200	0.231	0.312	0.972	0.531	0.824
employees	229	−0.046*	0.061*	0.409	0.241	0.352	0.934	0.552	0.829

Note: values in the table – regression coefficients significant at p = 0.05 – hypothesis accepted; coefficients marked with * significant at p > 0.05 – hypothesis rejected;

The fit indices of the SEM model in different subgroups are presented in Table 5. It should be noted that the ratios χ^2/df range from 2.4 to 3.9 and are at an acceptable level; RMSEA – around the acceptable level or above accordingly to different studies [15]; NFI and CFI are close to the acceptable level (≥ 0.9, [15]); Akaike Information Criterion (AIC) suggests that the model best fits to the population of all respondents (the lowest value of this criterion).

Table 5. Fit indices of the SEM model in different subgroups.

Group\Index	χ^2	χ^2/df	RMSEA	NFI	CFI	AIC
All	701.93	3.9	0.100	0.821	0.860	2.655
Age						
≤ 30	449.94	2.5	0.091	0.806	0.873	2.835
>30	490.91	2.7	0.110	0.746	0.820	5.523
Work						
Non related to ICT	453.3	2.5	0.107	0.762	0.839	4.811
Related to ICT	489.4	2.7	0.096	0.795	0.859	3.169
Status						
Students	425.81	2.4	0.087	0.795	0.869	2.782
Employees	609.76	3.4	0.105	0.805	0.853	3.113

Note: degree of freedom (df) = 181; p for χ^2 = 0.000; RMSEA – Root Mean Square Error of Approximation; NFI – Normed Fit Index; CFI – Comparative Fit Index; AIC - Akaike Information Criterion

5 Discussion and Conclusion

The study results provide an interesting insight into the perception of mobile technology usability for knowledge providing by different types of target users.

Taking into account the respondents' age, it was interesting to discover that all the hypotheses of the adjusted TAM model were supported in the group of younger respondents. However, for older respondents, the hypothesis that information quality might influence the perceived usefulness, and that the information navigation properties might affect the perceived ease of use were not supported. These findings are in line with other investigations concerning the differences in ICT perception by younger and older people [27]. For younger generations, the quality of mobile interface might be sufficient to absorb and navigate the mobile information whereas older people might experience difficulties with the perception of mobile information displayed on mobile devices; the information quality in terms of mobile device interface (layout, small characters) might not be sufficient for them and the navigation interface (lack of a keyboard/mouse) might impede information acquisition. Besides, the information quality in terms of language might not be satisfied for older generations, who are used to a more formal style of information delivery as opposed to short, concise, and informal style required by mobile devices to which younger people are used to.

In relation to activities/work being performed by respondents, it was interesting to discover similar results in the group of users non-related to ICT with older respondents. It seems that having professional background related to ICT gives a similar advantage to a user as being young and people that do not have an ICT background might experience similar difficulties as older respondents.

Concerning the third angle of investigation – respondents status (students versus employees) – our results are in line with the previous findings where all hypotheses have been confirmed in the group of younger users. As far as the employees are concerned, the relations between information access and perceived usefulness, as well as information quality and perceived usefulness were not confirmed (just as in the case of the whole population). It should be noted, however, that the population of employees has been changing and current students will enter labor market soon. Thus, the labor market will be enriched by individuals who perceive mobile technology and its usability in a quite different way compared with those already working. Since the characteristics of employees will change, the work environment should also change to meet the expectations of younger generations.

6 Implications for Research and Practice

The study, and especially the adjusted TAM model, might form the basis for further investigation into the perception of the usability of mobile technology for knowledge providing for different target users.

The research results might be useful also for practitioners by paying attention to the fact that the population of mobile device users is highly diversified. Hence, the content intended to be presented on mobile devices should be customized to a target user and should take into account the difference in age, professional background, and current activities being performed.

7   Limitations and Future Directions

The study results of testing the TAM model in different subgroups of respondents should be taken in caution. After dividing the dataset, we received in most cases the smaller than usually recommended (200) population of respondents in each of the subgroups. In future research we would like to extend our measurement model and the questionnaire by adding more items that assess the structural model constructs and conduct a more extensive survey in different groups of respondents.

Funding

This research has been financed by the funds granted to the Faculty of Management, Cracow University of Economics, Poland, within the subsidy for maintaining research potential.

References

1. Ally, M., Prieto-Blázquez, J.: What is the future of mobile learning in education? Int. J. Educ. Technol. High. Educ. **11**(1), 142–151 (2014)
2. Anderson, J.C., Gerbing, D.W.: Structural equation modeling in practice: a review and recommended two-step approach. Psychol. Bull. **103**(3), 411–423 (1988)
3. Barata, J., Rupino Da Cunha, P., Stal, J.: Mobile supply chain management in the Industry 4.0 era: an annotated bibliography and guide for future research. J. Enterp. Inf. Manag. **31** (1), 173–192 (2018). https://doi.org/10.1108/JEIM-09-2016-0156
4. Davis, F.D.: Perceived usefulness, perceived ease of use, and user acceptance of information technology. MIS Q. **13**, 319–340 (1989)
5. Davis, F.D., Bagozzi, R.P., Warshaw, P.R.: User acceptance of computer technology: a comparison of two theoretical models. Manag. Sci. **35**(8), 982–1003 (1989)
6. Davison, C.B., Lazaros, E.J.: Adopting mobile technology in the higher education classroom. J. Technol. Stud. **41**(1), 30–39 (2015)
7. De la Fuente, A., Ciccone, A.: Human capital in a global and knowledge-based economy, vol. 562. Office for Official Publications of the European Communities (2003)
8. Fishbein, M., Ajzen, I.: Belief, Attitude, Intention and Behavior: An Introduction to Theory and Research. Addison-Wesley, Reading (1975)
9. Foray, D., Lundvall, B.: The knowledge-based economy: from the economics of knowledge to the learning economy. In: The Economic Impact of Knowledge, pp. 115–121 (1998)
10. Gandhewar, N., Sheikh, R.: Google android: an emerging software plat-form for mobile devices. Int. J. Comput. Sci. Eng. **1**(1), 12–17 (2010)
11. Hair Jr., J.F., Black, W.C., Babin, B.J., Anderson, R.E., Tatham, R.L.: Multivariate Data Analysis, 6th edn. Prentice-Hall, Upper Saddle River (2006)
12. Hampton, T.: Recent advances in mobile technology benefit global health, research, and care. JAMA **307**(19), 2013–2014 (2012)
13. Harris, R.G.: The knowledge-based economy: intellectual origins and new economic perspectives. Int. J. Manag. Rev. **3**, 21–40 (2001)
14. Heflin, H., Shewmaker, J., Nguyen, J.: Impact of mobile technology on student attitudes, engagement, and learning. Comput. Educ. **107**, 91–99 (2017)
15. Hooper, D., Coughlan, J., Mullen, M.R.: Structural equation modelling: guidelines for determining model fit. Electron. J. Bus. Res. Methods **6**(1), 53–60 (2008)

16. Hwang, G.J., Wu, P.H.: Applications, impacts and trends of mobile technology-enhanced learning: a review of 2008–2012 publications in selected SSCI journals. Int. J. Mob. Learn. Org. **8**(2), 83–95 (2014)
17. Junglas, I., Abraham, C., Ives, B.: Mobile technology at the frontlines of patient care: understanding fit and human drives in utilization decisions and performance. Decis. Support Syst. **46**(3), 634–647 (2009)
18. Korres, G.M.: Technical Change and Economic Growth: Inside the Knowledge Based Economy. Routledge, Abingdon (2016)
19. Lundvall, B.A., et al.: The New Knowledge Economy in Europe: A Strategy for International Competitiveness and Social Cohesion. Edward Elgar Publishing, Cheltenham (2002)
20. Martin, F., Ertzberger, J.: Here and now mobile learning: an experimental study on the use of mobile technology. Comput. Educ. **68**, 76–85 (2013)
21. Mehdipour, Y., Zerehkafi, H.: Mobile learning for education: benefits and challenges. Int. J. Comput. Eng. Res. **3**(6), 93–101 (2013)
22. Prusak, L.: Knowledge in Organisations. Routledge, Abingdon (2009)
23. Rowles, D.: Mobile Marketing: How Mobile Technology is Revolutionizing Marketing, Communications and Advertising. Kogan Page Publishers, London (2017)
24. Sekaran, U., Bougie, R.: Research Methods for Business: A Skill Building Approach. Wiley, Hoboken (2016)
25. Soja, E., Soja, P., Paliwoda-Pękosz, G.: Solving problems during an enterprise system adoption: does employees' age matter? In: Wrycza, S. (ed.) SIGSAND/PLAIS 2016. LNBIP, vol. 264, pp. 131–143. Springer, Cham (2016). https://doi.org/10.1007/978-3-319-46642-2_9
26. Soja, E., Soja, P.: Exploring root problems in enterprise system adoption from an employee age perspective: a people-process-technology framework. Inf. Syst. Manag. **34**(4), 333–346 (2017)
27. Soja, E.: Information and communication technology in active and healthy ageing: exploring risks from multi-generation perspective. Inf. Syst. Manag. **34**(4), 320–332 (2017)
28. Stal, J., Paliwoda-Pękosz, G.: Mobile technology in knowledge acquisition: a preliminary study. In: Ulman, P., Węgrzyn, R., Wójtowicz, P. (eds.) Knowledge, Economy, Society: Challenges and Tools of Modern Finance and Information Technology, pp. 123–132. Foundation of the Cracow University of Economics, Cracow (2017)
29. Stal, J., Paliwoda-Pękosz, G.: Towards integration of mobile technology and knowledge management in organizations: a preliminary study. In: Kowal, J. et al. (ed.) Innovations for Human Development in Transition Economies. Proceedings of the International Conference on ICT Management for Global Competitiveness and Economic Growth in Emerging Economies, Wrocław, Poland, pp. 204–214 (2017)
30. Stal, J., Paliwoda-Pękosz, G.: Why M-learning might appeal to organisations? In: Themistocleous, M., Morabito, V., Ghoneim, A. (eds.) Proceedings of the 13th European, Mediterranean & Middle Eastern Conference on Information Systems, pp. 139–148. Cracow University of Economics, Kraków (2016)
31. Stal, J.: Data personalization in mobile environment: the content adaptation problem. In: Tvrdíková, M., Ministr, J., Rozehnal, P. (eds.) Proceedings of the 14th International Conference on Information Technology for Practice, pp. 181–188. Technical University of Ostrava (2011)
32. World Bank: Building Knowledge Economies: Advanced Strategies for Development. World Bank (2007)
33. Wu, J.H., Wang, S.C.: What drives mobile commerce?: an empirical evaluation of the revised technology acceptance model. Inf. Manag. **42**(5), 719–729 (2005)
34. Yousafzai, S.Y., Foxall, G.R., Pallister, J.G.: Technology acceptance: a meta-analysis of the TAM: Part 1. J. Model. Manag. **2**(3), 251–280 (2007)

Determinants of Academic E-Learning Systems Acceptance

Stanislaw Wrycza and Michal Kuciapski[✉] [iD]

Department of Business Informatics, University of Gdansk, Jana Bażyńskiego 8,
80-309 Gdańsk, Poland
{swrycza, m.kuciapski}@ug.edu.pl

Abstract. From the beginning of the e-learning technology era, many concerns have been raised regarding the use of e-learning systems in everyday academic didactics. Thus, a number of relevant studies have been conducted in the field of e-learning acceptance, mostly with the help of the Unified Theory of Acceptance and Use of Technology (UTAUT). This study investigates factors which influence the acceptance of academic e-learning technologies with the use of the modified UTAUT model. The basic UTAUT model was extended by new factors under examination: system interactivity (SIN) and the area of scientific expertise (ASE). In the survey, a total number of 242 academic teachers were asked to fill out the UTAUT-formatted questionnaire to determine their intention to use e-learning. Therefore, the paper contributes to technology acceptance theory, applied in e-learning, by extending the UTAUT model with two new variables—ASE and SIN—and by verifying the model validating their useful-ness. Study results are also valuable for practitioners, such as e-learning systems designers and developers. Factors such as performance expectancy, system interactivity, and area of scientific expertise were crucial for e-learning system development, to ensure that such an information system is widely accepted by end users such as faculty.

Keywords: Information systems · E-learning · UTAUT
Technology acceptance · Higher education · Faculty

1 Introduction

E-learning technologies offer an enjoyable training environment due to the presentation of the materials in various forms such as: video, audio, animation or simulations. Furthermore, through the elaboration of web 2.0 technologies, technological tools offer much greater social interaction. The effect pointed out by Fidani and Idrizi [1] is that e-learning technologies have proven to significantly boost the students' and teachers' performance and productivity. In the context of e-learning success and popularity, surprisingly many educators are reluctant to accept and use this new technology in their teaching [2]. Therefore, there is a constant need to search for the causes of such a state, which is the aim of this paper. Received feedback should have an impact on a sub-stantial expansion of interest in e-learning and its applications among academic staff.

© Springer Nature Switzerland AG 2018
S. Wrycza and J. Maślankowski (Eds.): SIGSAND/PLAIS 2018, LNBIP 333, pp. 68–85, 2018.
https://doi.org/10.1007/978-3-030-00060-8_6

Many researchers have proposed theories and models of technology acceptance in order to explain and predict users' acceptance or adoption of particular technologies [3], outlining profiles and differences [4]. This crucial challenge has inspired many researchers to propose various models for its acceptance and adoption by potential users. Many IS scholars, starting from Fishbein and Ajzen [5] and Davis [6], have proposed and verified their theories, solutions, models, and methods for the acceptance, adoption and usage of technology. In particular, the following approaches should be mentioned, among others: TAM—Technology Acceptance Model [7], IDT—Innovation Diffusion Theory [8], TAM 2 [9], UTAUT—Unified Theory of Acceptance and Use of Technology [10], TAM 3 [11] and UTAUT2 [12], to name the most significant ones. The TRA first explains the drive of human actions with two constructs: attitudes toward target behavior (ATB) and attitudes toward subjective norm (SN). The causal relationship between attitudes and actions suggested by TRA is rather strong. The Motivational Model distinguishes the effects of extrinsic (EM) and intrinsic motivation (IM) in influencing the level of technology acceptance. The TPB, based on the viewpoint of TRA, includes perceived behavioral control (PBC) to explain the relationship between attitudes and behaviours. While the C-TAM-TPB model combines constructs of TAM and TPB, the MPCU focuses on external factors that might influence the acceptance level, such as job fit (JF) and resources (RES) available for using the technology. TAM 2 addresses technology acceptance in mandatory settings by refocusing on the effects of subjective norms (or social influences). TAM 3 includes new external variables (determinants): computer self-efficiency (CSE), computer Playfulness (CP), perceived enjoyment (PE), and objective usability (OU).

Among various technology acceptance models, it is UTAUT that has attracted particular attention in the area of acceptance research from its very inception [13, 14]. UTAUT has condensed the 32 variables found in the existing eight models (TRA, TPB, TAM, MM, C-TPB-TAM, MPCU, IDT and SCT) into four main variables and four moderating factors. The UTAUT model [10] enables the assessment of dependent variables (DV), i.e.—behavioural intention (BI), and then usage behaviour (UB), by estimating the influence of four key independent variables (IV): performance expectancy (PE), effort expectancy (EE), social influence (SI), and facilitating conditions (FC). Besides, UTAUT considers the influence of four moderators (M): age, gender, prior experience with technologies, and voluntariness of use, on the significance of independent variables impacting BI.

To solve the problem of determining variables impacting e-learning acceptance by faculty, the UTAUT model was chosen as an adequate tool. It turned out that, while TAM is capable of predicting technology adoption success at an accuracy rate of 30% and TAM2 (TAM extension) at 40% [13], the combination of the variables and moderators in the UTAUT model have increased the prediction rate to 70% [15]. Moreover, UTAUT has been the tool used in acceptance surveys regarding various technological contexts, such as: e-commerce, internet banking, e-government and social networking and e-learning.

The form of the UTAUT model was modified by a number of researchers, by reshaping the UTAUT in terms of extending it with some additional independent variables and determinants and/or omitting certain classical determinants and moderators. Some examples of modifications regarding new variables or determinants may include: system flexibility [16], actual usage [4], and many others presented in the second part of the paper. Modifications included both new independent and external variables. Independent variables are the ones that have direct impact on BI, such as attitude towards use (ATU) proven in [28]. New external variables have an indirect impact on BI, such as system enjoyment (SE) having an impact on performance expectancy (PE) that is connected with BI, as described in Alrawashdeh et al. [16]. This highlights the fact that the UTAUT model can be significantly modified to construct new and better technology acceptance models. This is especially important, as many studies confirmed that, in different countries for miscellaneous technologies and various target groups such as students, employees, and academicians, technology acceptance determinants can be different, as pointed out in [17–19]. Accordingly, we should agree with Marchewka et al. [20] that, despite the high recognition of acceptance models like UTAUT, their validity, in the context of information systems, requires further testing.

This study was connected with developing the technology acceptance model of e-learning with new variables, with regard to an under-researched target group of faculty, in Poland, where there have been no studies on e-learning acceptance and few about technology acceptance at all. As the elaborated model was based on UTAUT, its validity was also tested. The paper comprises of five sections. After the introduction, the second section contains a comparative analysis of publications regarding the different acceptance models applied to e-learning. In the third section, the transformation from the classical UTAUT model, to our UTAUT-EL model, is carried out. The fourth section deals with the verification of the UTAUT-EL model in respect of ten research hypotheses presenting study results, that are also valuable for practitioners as information systems producers. The findings of the research are synthesized in the conclusion. The paper finishes with a presentation of study limitations and further research.

2 Theoretical Framework

Because of its significance for academic institutions, scholars, and students, e-learning and collaboration has been the subject of many surveys regarding its acceptance and adoption at universities [21]. The comparative analysis of academic e-learning technology acceptance, made by the authors, included over fifteen cases conducted, among others, in: Indonesia, South Korea and Spain. Ten criteria were employed for making such a comparison. Ten cases that introduce new elements, such as variables, connections, or moderators for acceptance theories are presented in Table 1. While the titles of certain columns express the contents clearly, some require further explanation. First and foremost, adjustments to the selected model variables and moderators are: IV—independent variables, DV—dependent variables, EV—external variables, M—moderators. Because of the qualitative nature of the specific questions in the UTAUT questionnaire, a reliable scale is required to assess the answers. Most of the researchers have used the 5-point Likert scale, but there are some exceptions, extending this scale. Some specific notation was used for hypotheses verification—a plus sign above arrows (e.g. PE $\xrightarrow{+}$ BI) means that the relationship between variables was significant, while a minus would indicate that it is insignificant. The comparative analysis of e-learning technology acceptance contained in Table 1 is a foundation to propose a new model for e-learning acceptance, according to the premises used, from the perspective of choosing the right basic acceptance model and research methodology.

Table 1. The comparative analysis of academic e-learning acceptance surveys.

1. Auth./ publ.	2. Country	3. Scope of research	4. Target group	5. Resp. no.	6. Accept. model	7. Model variables / moderators	8. Scale quest.	9. Data analysis methods applied	10. Hypothesis verification
[13]	Nigeria	ICT acceptance by academic staff at two universities	Academic staff of Nigerian Universities – proportional, from each department	98	Modified UTAUT	IV: UTAUT standard plus Anxiety (AX), Attitude towards Use (ATUT), Self-efficiency (SE) DV: UTAUT standard M: UTAUT standard	Likert (1-5)	Descriptive statistics methods, Cronbach's Alpha, coefficient of determination, significance test	PE ●—+→ BI EE ●—+→ BI ATUT ●—+→ BI SI ●—−→ BI FC ●—−→ BI SE ●—−→ BI AX ●—−→ BI
[22]	Pakistan	The acceptance of e-learning technology	Staff from MBA/MPA, English and computer departments from Pakistani University	70	TAM	IV: Facilitating conditions (FC), Perceived ease of use (PEOU), Computer efficiency (CE), Perceived usefulness (PU) DV: Intention to Use	Likert (1-5)	Descriptive statistics methods, regression equation, beta, significance test	PEOU ●—+→ ITU PU ●—+→ ITU CE ●—+→ ITU FC ●—+→ ITU (much lower than others)
[23]	Indonesia	Blog technology acceptance in teaching and learning	Students from Indonesian University	49	Modified UTAUT	IV: UTAUT standard minus Facilitating conditions (FC) DV: UTAUT standard plus Level of Actual Usage (LAU) M: Gender, Experience	Likert (1-6)	Descriptive statistics methods, Cronbach's Alpha, standardize β-Coefficients, significance test, R^2	PE ●—+→ BI EE ●—−→ BI SI ●—+→ BI BI ●—−→ LAU Gender and experience have no impact on BI
[24]	Taiwan	Cross-level analysis of factors of e-learning behavioral intention (group-level and individual-level)	Various respondents from Taiwan	932	Modified UTAUT	IV: (a) individual-level: Performance expectancy (PE), Effort expectancy (PE), Perceived behavioral control (PBC) (b) group-level: Incentive (I), Manager influence (MI), Colleague influence (CI) DV: UTAUT standard	Likert (1-5)	Descriptive statistics methods, Cronbach's Alpha, coefficient of determination, beta, Variance Inflation Factor (VIF) measuring the severity of multicollinearity, significance test, t –value, γ coefficient	PE ●—+→ BI EE ●—+→ BI PBC ●—+→ BI CI ●—+→ BI MI ●—+→ BI I ●—+→ MI I ●—+→ CI
[16]	Jordan	Acceptance of web-based training system	Jordanian public sector employees	290	Modified UTAUT	IV: UTAUT standard plus System flexibility (SF), System enjoyment (SE) EV: System enjoyment (SE) DV: UTAUT standard	Likert (1-7)	Descriptive statistics methods, Structural Equation Model (SEM), Cronbach's Alpha, coefficient of determination, beta, validity test, composite reliability, average variance extracted, convergent validity, Chi-square to degrees of freedom	PE ●—+→ BI EE ●—+→ BI SI ●—+→ BI FC ●—+→ BI SF ●—+→ BI SE ●—+→ BI SE ●—+→ PE SE ●—+→ EE EE ●—+→ PE FC ●—+→ EE SF ●—+→ PE
[25]	Spain	Gender influence on acceptance and adoption of e-learning platforms by students	Students of a main University from the south of Spain	189 (66 males and 123 females)	Modified TAM, TAM2 and TAM3	IV: perceived usefulness (PU), perceived ease of use (PEOU) EV: result demonstrability (RES), perception of external Control (PCE), perceived enjoy (ENJ) DV: Behavioral intention, Use of the Platform e-Learning (USE) M: Gender	Likert (1-5)	Descriptive statistics methods, PLS, average variance extracted (AVE), Structural Equation Modeling (SEM), Cronbach's α coefficient, path coefficients (β), R^2, p-values, regression coefficients	BI ●—+→ USE ENJ ●—+→ PEOU PCE ●—+→ PEOU PEOU ●—+→ ITU PEOU ●—+→ PU PU ●—+→ ITU RES ●—+→ PU no differences between males and females

[1]	Macedonia	Students' acceptance of a LMS in university education	Students of University in Macedonia	213	Modied UTAUT	IV: UTAUT standard, Attitude towards using technology (ATUT) DV: UTAUT standard	Likert (1-5)	Descriptive statistics methods, Cronbach's alpha, KMO and Bartlett's test of sphericity, AVE, χ2/d.f., GFI, RMSEA, Root Mean Square Residual (RMR), Normed Fit Index) (NFI), Non-normed Fit Index) (NNFI)	PE ●—+→ BI PE ●—→ ATUT EE ●—+→ PE (very significant) EE ●—-→ ATUT EE ●—+→ BI ATUT ●—+→ BI SI ●—+→ ATUT SI ●—+→ BI FC ●—+→ BI FC ●—+→ EE
[26]	Iran	E-learning acceptance in teaching English language	Students and teachers of Iranian University	103 - 90 stud. and 13 teach.	TAM	IV: TAM standard DV: TAM standard	Likert (1-5)	Descriptive statistics methods, Cronbach's Alpha, Pearson Correlation, significance test	PU ●—+→ ITU PEOU ●—→ ITU PEOU ●—+→ PU
[27]	Finland	Teachers acceptance of continuing e-learning systems usage	University educators from Finnish University that use Moodle platform	175	Modified UTAUT	IV: Access (ACCESS), Perceived behavioral control (PBC) and Compatibility (C); Performance expectancy and Perceived usefulness; Effort expectancy and Perceived ease of use; Social influence DV: Continuance Intention (CI)	Likert (1-5)	Descriptive statistics methods, PLS, Composite reliability (CR), beta, significance test	PU ●—+→ CI ACCESS ●—+→ CI C ●—-→ CI PBC ●—+→ CI PEOU ●—+→ CI SI ●—+→ CI
[28]	South Korea	Intrinsic motivators and extrinsic motivators in promoting e-learning in the workplace	Employees of mid-size food service company in South Korea	226	Modified UTAUT	IV: a) IM - Intrinsic motivators (Effort expectancy, Attitude toward use, and Anxiety) b) EV - Extrinsic motivators (PE, SI and FC)	Likert (1-7)	Descriptive statistics methods, Cronbach's alpha coefficient, confirmatory factor analysis (CFA), structural equation modeling (SEM)	IM ●—+→ BI EM ●—-→ BI EM ●—+→ IM

The comparative analysis in Table 1 confirms that UTAUT has become a popular tool for technology and software research all over the world, in countries from different continents. Modifications of the classical UTAUT model involve supplementing it with additional independent variables and determinants and/or omitting some classical determinants and moderators. The next criterion—the target group—revealed that acceptance surveys, involving questionnaires, were conducted among teachers, students, and employees. The comparison in column 5 reveals that various number of respondents of the survey, from 49 to 932, with an average of about 120. Independently of the number of respondents, the UTAUT or TAM models were applied as the basic acceptance approaches. As column 6 states, the acceptance research studies were accomplished, first and foremost, by using the modified UTAUT model.

All authors used research methodology well-established for studying technology acceptance, based on questionnaires, consisting of two to four assertion statements (items) per variable measured by the Likert scale. The form of the questions was closely related to typical UTAUT [10] or TAM surveys [6]. For example, items for performance expectancy variable of e-learning technology acceptance in workplace have been defined in Sun et al. [28] as:

1. I would find e-learning useful in my job.
2. Using e-learning enables me to accomplish tasks more quickly.
3. Using e-learning increases my productivity.
4. If I use e-learning, I will increase my chances of getting a raise.

In the papers dedicated to technology acceptance (Table 1) there many new independent variables are proposed, such as: anxiety (AX), self-efficiency (SE), perception of external control (PCE), computer efficiency (CE), IM (intrinsic motivators), and many others. External variables (EV) are a new type in relation to the classical UTAUT model. They are identified as independent variables, that have an indirect impact on dependent variables (DV) through direct impact on other independent variables (IV).

Not all classical UTAUT independent variables were used in the broad range of the modified models. According to Table 1, the researchers quite often decided against including two determinants: social influence (SI) and facilitating conditions (FC). Furthermore, the moderators, in general, are omitted by the authors in their modified UTAUT models. Only a few models included two of four UTUAT moderators—gender and experience. While, in the original UTAUT model, nine hypotheses were verified, the authors of the modified models formulated and verified from three to ten hypotheses. The last column of Table 1 includes the results of hypothesis verification with the proposed, specific, and straight notation. Detail interpretation of particular variables, items for them, and connections, is included in particular papers where their references are given in first column.

Conducted analysis of subject matter literature, presented synthetically in Table 1 allowed information to be gained about existing variables, and connections in technology acceptance models, especially related to e-learning. This facilitated concentration on the elaboration of new determinants, that would contribute to general technology acceptance theories.

3 Research Model

The aim of the comparative analysis of papers related to e-learning acceptance was to review and assess independent and external variables, moderators, and hypotheses which refer to this academic field of activity. Analysis was carried out with respect to the authors' own experience in using e-learning technologies in day-to-day academic practice and being members of distance education council. This proves that there is still a necessity to search for new factors that influence the intention to use e-learning by faculty for university teaching. In the process of analyzing the determinants for the successful implementation and acceptance of this form of academic teaching, a new model was proposed in this study, called UTAUT-EL. The authors proposed the modified UTAUT model, containing all classical UTAUT model variables: performance expectancy (PE), effort expectancy (EE), social influence (SI), facilitating conditions (FC), and behavioural intention to use (BI), as well as two new ones—system interactivity (SIN) and area of scientific expertise (ASE) (Fig. 1).

PE, included in the model and existing in UTAUT, measures the degree to which an individual believes that using e-learning will help him or her attain gains in job performance. Improving a student's performance expectancy towards blog technology usage was essential to the student's level of intent to adopt a blog for learning [23]. EE, in the model, indicates the perceived ease of utilizing e-learning technology in the faculty. SI measures the degree to which an individual perceives that it is important others believe he or she should use e-learning. For e-learning platforms usage, the faculty board can encourage academic teachers to use e-learning systems to teach students. FC, in the model, reports support from the organization to adopt e-learning successfully for work, perceived by the faculty. As presented in UTAUT, FC is connected only with use behaviour and thus does not have impact on behaviour intention to use technology (Fig. 1). BI variable in the model represents intention to use e-learning technologies for teaching by faculty.

Taking into account the literature review, the results presented above, and their long-term experience in e-learning, the authors selected the set of variables and moderators during brainstorming sessions. Consequently, the proposed research model includes two new original ones: system interactivity (SIN) and area of scientific expertise (ASE).

Various researchers confirm that technical aspects of using e-learning strongly influence its acceptance. Computer efficiency (CE) is proved by Waheed [22] to have an impact on acceptance of e-learning technology by faculty. CE expresses the user's previous experience and command on computer usage in order to perform his or her job effectively [29]. Studies of [16, 24, 25, 30] highlight that e-learning acceptance is strongly correlated with interaction with e-learning systems. Perceived behavioural control (PBC) was positively verified by Liao et al. [24] to significantly influence behavioural intention to use e-learning. PBC refers to people's perceptions of their ability to perform a given behaviour with system [31]. System flexibility in the study of Alrawashdeh et al. [16] is defined as the degree to which individual believes that he/she can access the system from anywhere, at any time, which has also been confirmed to significantly influence the acceptance of e-learning training systems by employees. Finally, result demonstrability (RES), understood as ease in presenting the results of work, was proved to be an important factor impacting on behavioural intention of using e-learning by students [25]. As a result of the integration of highlighted determinants, a more complex SIN variable was considered by authors as a potentially important factor in e-learning acceptance, regarding ICT technology features, functionalities, and user interfaces. It is understood as the perceived responsiveness, flexibility of use, and intuitiveness of e-learning technologies for its users. It means the efficiency of interaction with an e-learning system to execute tasks in various environments and conditions.

The second new independent variable—ASE, results from the fact that the research was conducted among all departments of the university. Certain specializations and faculties, such as IT, are currently using ICT technology in their everyday didactic and scientific work, while scholars of other specializations use e-learning very rarely or not at all. Therefore, they require extra training to achieve fluency in online teaching. Moreover, it seems highly probably that lecturers expect e-learning systems to match their faculty specifics. Unstructured interviews with 32 academics, from various faculties, proved that they expect e-learning solutions, such as e-learning platforms, to offer a wider spectrum of tools that are dedicated towards different faculties, such as business and administration, computer science, or connected with teaching foreign

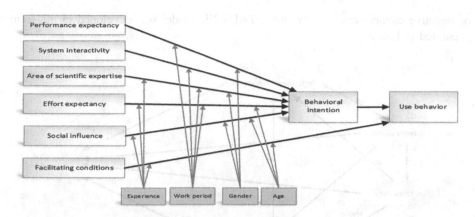

Fig. 1. Preliminary version of the modified UTAUT model of academic e-learning acceptance.

languages. Inclusion of ASE, as a technology acceptance variable in the proposed model, is supported by the study results of [32] where the authors point out that: 'professional development programs should assess teachers' digital learning motivation profiles and design professional learning experiences that expand upon teachers' beliefs, values and attitudes, and the conceptual themes of most importance to them.' Therefore, variable ASE has been included in the proposed model. ASE is defined as the degree to which an academic teacher believes that using e-learning technologies both fits with and improves his/her teaching within their specific academic field. Technology acceptance studies, directly related to proposed ASE variables, have not been found in subject matter literature. However, there are studies indirectly justifying inclusion of this variable in the model. A study of Pardamean and Susanto [23] proves that the level of actual usage (LAU) variable influences students' acceptance of blog technology in teaching and learning. LAU is explained as the current level of technology utilization, due to the positive feedback on the solution functionality. Furthermore, the Compatibility (C) variable influences teachers' acceptance of continuing e-learning systems usage has been positively verified by Islam [27]. Compatibility is defined as technology or system compatibility with users' needs and style of work [33]. As a result of proposing SIN and ASE variables, a developed model—called UTAUT-EL—took its preliminary form as shown in Fig. 1.

The independent variables, and moderators influencing behavioural intention, were statistically verified under the criterion of the significance of the relationships between them. This analyses, evaluations, and interpretations of the final relationships were conducted and received with the help of Structural Equation Modelling (SEM). SEM was based on continuous fit confirmatory factor analyses (CFAs), which excluded:

- observed variables whose impact on unobserved variables was lower than 0.5;
- observed variables, where residual covariance was higher than 1.4;
- connections between unobserved variables, where the relationship was not significant.

As a consequence, the following determinants were excluded from the preliminary version of the modified UTAUT model: social influence and facilitating conditions. The moderators were removed from the preliminary model version, due to the existence of negative covariances. Finally, the UTAUT-EL model was developed in the form presented in Fig. 2.

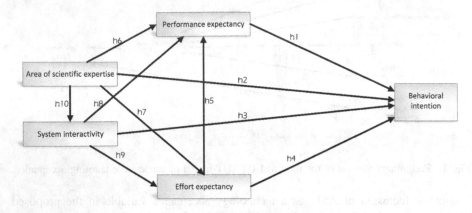

Fig. 2. UTAUT-EL model.

The relationships are equivalent to ten research hypotheses, indicated in Fig. 2 and formulated in Table 2.

Table 2. Research hypotheses.

Hypothesis number	Connection	Description
H1	PE ●⁺→ BI	Performance expectancy has direct effect on the intention to use e-learning technologies.
H2	ASE ●⁺→ BI	Area of scientific expertise has direct effect on the intention to use e-learning technologies.
H3	SIN ●⁺→ BI	System interactivity has direct effect on the intention to use e-learning technologies.
H4	EE ●⁺→ BI	Effort expectancy has direct effect on the intention to use e-learning technologies.
H5	EE ●⁺→ PE	Effort expectancy has a direct impact on perceived performance expectancy.
H6	ASE ●⁺→ PE	Area of scientific expertise has a direct impact on perceived performance expectancy.
H7	ASE ●⁺→ EE	Area of scientific expertise has a direct impact on perceived effort expectancy.
H8	SIN ●⁺→ PE	System interactivity has a direct impact on perceived performance expectancy.
H9	SIN ●⁺→ EE	System interactivity has a direct impact on perceived effort expectancy.
H10	ASE ●⁺→ SIN	Area of scientific expertise has a direct impact on perceived system interactivity.

4 Model Validation and Discussion

As stated previously, the study was based on a questionnaire survey using the CAWI (Computer-Assisted Web Interview) method. An invitation to take part in the survey was sent to 242 scholars via the University E-learning Platform. Data was collected twice, over a two-week period, at University of Gdansk in Poland. Eventually, 82 teachers from all ten faculties filled in the questionnaire (response rate of 34%). The first section included questions regarding the characteristics of the respondents. The crucial second part included 22 assertion statements and three to four items for each variable, expressed in the preliminary UTAUT model (Fig. 1). Each question was measured using the 5-point Likert scale. Items for variables were prepared according to the UTAUT approach, with a minor adaptation of assertion statements of classical UTAUT model variables (EE, PE, SI, FC, BI) to the field of e-learning technologies acceptance. They also had a very similar form to items presented in papers juxtaposed in Table 1. For new proposed variables of SIN and ASE, proper assertion statements have been prepared. For SIN variable 4, items have been elaborated:

1. E-learning tools' interfaces ensure fast and efficient realization of teaching.
2. E-learning applications' interfaces do not create problems for students during classes.
3. E-learning technologies allow teaching to be conducted in a more flexible manner than before.
4. Available functionality in e-learning tools meets my expectations.

ASE variable had assigned 3 items:

1. The specificity of my scientific specialization favours the use of e-learning technologies.
2. My expertise in the field of ICT (Information and Communication Technologies) encourages the use of e-learning technologies.
3. The use of e-learning technologies requires significant changes in my previous style of teaching.

Regression analysis was used to verify the influence of the four proposed variables of UTAUT-EL—PE, EE, SIN and ASE on BI (Fig. 2). The results were used to accept or reject the stated null hypotheses (Table 2). IBM SPSS Statistics 21 was applied while calculating the reliability coefficients and the explanatory factor analysis [34, 35]. SEM was applied in assessing interactively the direct and indirect effects between exogenous and endogenous variables of the complex model structure [36]. SEM was used, due to both its popularity and the fact that it has also been tried and tested in the field of technology acceptance. The advantage of SEM is that it considers both the evaluation of the measurement model, and the estimation of the structural coefficient, at the same time. Thus, it allows for dynamic model adaptation with constant fit confirmatory factor analysis. Such research methodology ensures the correctness of a given model.

Data validity tests showed that all 82 cases were valid. The validity was concerned with reducing the possibility of incorrect answers during the data collection period [37]. Inter-construct correlation coefficient estimates were examined along with a particular item's internal consistency reliability, by using Cronbach's alpha coefficient estimates [38, 39]. Table 3 includes the relevant results.

Table 3. Data reliability.

Variable	Cronbach's alpha based on standardized items
PE	0.876
EE	0.770
ASE	0.632
SIN	0.838
BI	0.857

Generally, as stated by Zhang and Sun [40], reliability numbers greater than 0.6 are considered acceptable in technology acceptance literature. All items exceeded the recommended level. The research instrument confirmed that the data is internally consistent and acceptable, with a total reliability equal to 0.854.

The final model of this study was obtained through a process including item exclusion. In order to assess the overall metric model, fit confirmatory factor analysis (CFAs) was performed. Consequently, the model meets all accuracy requirements of fit measures, as presented in Table 4.

Table 4. Fit indices of UTAUT-EL.

Fit indices	Recommended value	Result
$\chi2$/d.f.	<3	1.29
GFI (Goodness of Fit Index)	>0.8	0.887
RMSEA (Root Mean Square Error of Approximation)	<0.08	0.06
CFI (Comparative Fit Index)	>0.9	0.965
AGFI (Adjusted Goodness of Fit Index)	>0.8	0.813

Regression analysis was used for the ten stated hypotheses testing. Their confirmation was examined by path coefficients (standardized $\beta \geq 0.2$ were accepted), and their significance levels ($p < 0.05$ were accepted). Table 5 shows the overall results of the hypotheses' verification.

Table 5. Hypothesis verification results.

Hypothesis	Path	Standardized β-coefficients	Significance
H1	PE ●—+→ BI	0.434	0.031
H2	ASE ●—-→ BI	0.026	0.888
H3	SIN ●—+→ BI	0.353	0.029
H4	EE ●—-→ BI	-0.187	0.215
H5	EE ●—+→ PE	-0.270	0.048

Furthermore, as shown in Table 6, indirect effects were measured to explain behaviour intention, regarding the use of e-learning technologies by academics, in closer detail. This also allowed us to measure the total effects for determinants.

Table 6. Indirect effects between variables.

Variable	ASE	SIN	EE
BI	0.298	0.203	-0.117
SIN			
EE	0.038		
PE	0.070	-0.028	
BI	0.298	0.203	-0.117

The findings of this study revealed that most of the proposed relationships were accepted as statistically significant. Figure 3 explains the elaborated structural model with path coefficients (β), significance (p), and the adjusted coefficients of determination (R^2) scores.

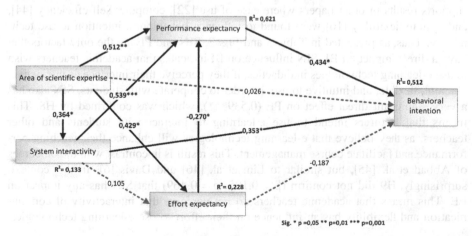

Fig. 3. UTAUT-EL—verification results.

Interestingly, the first hypothesis (H1) revealed that PE has a direct effect on the academic teachers' intention to use e-learning technologies. This hypothesis was accepted, since the statistical result revealed that there is a significant relationship between the PE and the lecturers' intention to use e-learning technology (0.434*) (Fig. 3). The meaning of the stars is as follows: ***—very significant impact, **—strongly significant impact, *—significant. This implies that academic teachers find e-learning technologies to be useful in improving their teaching performance, and as a way of expanding the channels of communication with students and other instructors. It confirms that academic teachers regard e-learning technologies as a tool that can extend and enhance the way they teach.

On the other hand, the relationship between EE and BI is not significant ($\beta = -0.187$, $p = 0.215$) (Table 5); hence, the second hypothesis (H2) is not supported. This result is consistent with findings in some previous studies, such as Sedana and Wijaya [18] and Dasgupda et al. [41]. In contrast, there are other studies that reported a significant influence between these two factors, such as Venkatesh et al. [10] and Chen et al. [42]. This indicates that the perceived level of effort required to use e-learning technologies by academics does not influence their decision as to whether or not to accept e-learning.

According to the results of the H2 verification, EE has no influence on behavioural intention; however, the confirmed H5 proves that EE has an indirect impact on the intention to use e-learning technologies (−0.117) (Table 6), by direct impact on PE (−0.270*). This confirms the findings of [7, 16, 43] who indicated that ease of use (similar to EE) affects usefulness (similar to PE) and user attitude. This highlights the fact that academic teachers also perceive the benefits gained from using e-learning technologies for tutoring from the perspective of the effort needed to be able to utilize them. The results in Table 6 point out that the higher the level of effort required, the lower the perceived PE.

In accordance with the confirmed third hypothesis (H3), this study provides evidence that SIN has a direct effect on the behavioral intention to use e-learning technologies (0.353*). As SIN condensed the variables of EOU, CSE and SF, it also supports results in other papers where ease of use [22], computer self-efficiency [44], and system flexibility [16] were found to influence behavioural intention to use technology. Thus, as presented in Table 5 and Fig. 3, SIN and PE are the only factors that have a direct impact on BI. SIN influence on BI indicates that academic teachers wish to use e-learning technologies in didactics, if they perceive their interface as interactive, efficient, flexible, and intuitive to use, in order to cooperate with students. SIN also has a very significant direct effect on PE (0.539***), which was confirmed by H8. This means that lecturers intend to use e-learning to interact with students and other teachers, as they believe that e-learning technologies will enhance their teaching performance and facilitate course management. This result is in contrast with the outcomes of Abbad et al. [45], but similar to Lim et al. [46] and Davis [6]. In this context, surprisingly, H9 did not confirm ($\beta = 0.105$; $p = 0.469$) that SIN has any impact on EE. This means that academic teachers do not perceive that interactivity of communication and flexibility has an influence on their effort to use e-learning technologies.

It is worth highlighting that SIN has the strongest total effect on BI (0.556). The influence of variables of EOU, CSE and SF, condensed in SIN, on EE was not examined in the subject matter literature.

The second relationship (H2), associated with elaborated variables, revealed that ASE does not have a direct effect on the academic teachers' intention to use e-learning technology, as the connection is not significant (β = 0.026; p = 0.888). The confirmed H2 does not mean that ASE does not have any influence on behavioural intention, as H6, H7 and H10 have been confirmed and showed that ASE has an important indirect impact on BI (0.298) (Table 6). As presented in Fig. 3, this is because of the direct relation between ASE and: PE (0.512**), EE (0.429*), as well as SIN (0.364*). In particular, accordingly to H6, academic teachers consider PE, gained by using e-learning technologies, as strongly related to a particular didactic or research field. This finding points to the demand for proper e-learning technology promotion among university employees, that is matched to the specifics of their particular faculties or even departments and is strictly connected to their area of tutoring or research. The same situation is confirmed for the relationship with EE (H7) and SIN (H10). Academic teachers perceive the level of EE, in using e-learning technologies and interactivity of communication, as dependent on a context of usage that directly corresponds to their respective fields of tutoring or research. It can be interpreted, that training of lecturers in the area of using e-learning technologies should be customized for particular faculties, departments, and research fields. As ASE is a new original variable, where its meaning is not connected to existing subject matter literature determinants, verification of the second hypothesis (H2) cannot be compared with other studies.

The model differs from the classical UTAUT model and its modifications. In contrast with UTAUT and many other studies, such as Chen et al. [42], it turned out that EE does not have a direct impact on BI. Furthermore, the study proved a connection between EE and PE does not exist in UTAUT. It is to be found in other studies, such as Sumak et al. [42] and Kuciapski [47]. As also proposed, SIN and ASE variables influence BI. We would agree with Marchewka et al. [20] that, despite the high recognition of technology acceptance models like UTAUT, their validity in the context of information systems requires further testing. The study also confirms what had been highlighted in the studies of [17–19] that cultural settings have an impact on technology acceptance, in this case on e-learning systems and tool acceptance by faculty for teaching.

Study results are valuable for practitioners, such as e-learning systems designers and developers. Research results point out which factors are important in e-learning systems, to have them widely accepted by users such as faculty. First of all, e-learning systems should be developed in such a way as to increase the performance of conducting didactics by faculty thanks to their use (Table 5). This can be interpreted as providing a wide spectrum of functionality for publishing didactic materials, communication, and assessment of students' progress in competences development, but with a significant level of interactivity with the system (Fig. 3). The effort required to learn e-learning solutions is not of much importance. Importantly, e-learning platforms should take into account the area of expertise of the faculty, by customizing functionality according to their scientific expertise (Table 6). Current solutions do not take this into account and do not offer a wide spectrum of tools that are dedicated for different faculties, such as business and administration, computer science, or connected with the teaching of foreign languages.

5 Conclusion

The authors investigated the behavioural intention of academic teachers to accept e-learning technologies such as education platforms, with the aim of advancing knowledge concerning factors which may influence their acceptance. This study utilized UTAUT as a base theory, while modifying it via the integration of additional constructs, such as system interactivity and the area of scientific expertise. Additional relationships between the variables were included in the new modified model called UTAUT-EL. Thus, the research contributed to the development of a context specific approach and the verification of the UTAT model.

This study enables several conclusions to be drawn. Firstly, it showed that academic teachers intend to use e-learning technologies mainly to improve their efficiency in teaching, since performance expectancy has the strongest direct effect on lecturers' intention to use e-learning technologies. Additionally, system interactivity has a significant impact on the intention to use e-learning technologies, with a total effect even greater than PE. This means that e-learning technologies designers should ensure that the system's components are highly interactive and intuitive to use and allow for the use of e-learning anywhere at any time. This may also be interpreted as an indication that users of e-learning platforms have a strong and defined control over the system. Academic teachers would also probably prefer to customize e-learning environments.

The newly included variable—the area of scientific expertise—does not have a direct impact on the intention to use e-learning technologies. But the research confirmed hypotheses that ASE directly affects performance expectancy, effort expectancy and system interactivity, and thus indirectly influences behavioural intention to use e-learning technologies. The study outcomes show that there is a need to analyse academic teachers' requirements for e-learning technologies, in respect of lecturers didactic or research specializations.

Study results are valuable for practitioners working in companies developing e-learning systems. Performance of conducting activities in e-learning systems, high interactivity with them, and adaptation of their functionality, according to the area of scientific expertise of the faculty concerned should be taken into account during their design and development. Current e-learning solutions do not offer various spectrums of tools that are dedicated for different faculties.

This study has some limitations. First, the value of R^2 of the elaborated model is 51.2% (Fig. 3). This means that behavioural intention to use e-learning technologies by faculty is explained in 51.2% of cases. Such a relatively high value shows that there is still a need to search for new technology acceptance variables and connections between them. The second limitation of this study is that data for new variables of system interactivity (SIN) and area of scientific expertise (ASE) was collected at one university in one country—which is a common practice but might be undermined in further studies. In addition, the meaning of ASE variables is strictly connected to the field of higher education; thus, it cannot be used for developing general purpose technology acceptance models.

Further studies will continue to confirm that the SIN and ASE variables have significant direct and/or indirect impact on behavioural intention to use e-learning technology in other countries. Moreover, those variables will be checked in different contexts (e.g. ERP systems, Office tools) and target groups (e.g. students, employees) to examine their importance for the research field of technology acceptance. It is especially important that the ASE variable will be adapted as a more general one—area of professional expertise (APE). APE will be examined, as a new variable, useful for the development of acceptance models for various kinds of technologies, and will therefore extend technology acceptance theories.

References

1. Fidani, A., Idrizi, F.: Investigating students' acceptance of a learning management system in university education: a structural equation modeling approach. In: ICT Innovations, pp. 191–200 (2012)
2. Kim, M.K.: Factors influencing the acceptance of e-Learning courses for mainstream faculty in higher institutions. Int. J. Instr. Technol. Distance Learn. 5(2), 111–116 (2008)
3. Barkhordari, M., Nourollah, Z., Mashayekhi, Z.: Factors influencing adoption of e-payment systems: an empirical study on Iranian customers. Inf. Syst. e-Bus. Manag. 15(1), 89–116 (2017)
4. Villani, D., et al.: Students' acceptance of Tablet PCs in Italian high schools: profiles and differences. Br. J. Educ. Technol. 49(3), 533–544 (2018)
5. Fishbein, M., Ajzen, I.: Belief, Attitude, Intention, and Behaviour: An Introduction to Theory and Research. Addison-Wesley, Reading (1975)
6. Davis, F.D.: Perceived usefulness, perceived ease of use, and user acceptance of information technology. MIS Q. 13(3), 319–340 (1989)
7. Davis, F.D., Bagozzi, R.P., Warshaw, P.R.: User acceptance of computer technology: a comparison of two theoretical models. Manag. Sci. 35, 982–1003 (1989)
8. Moore, G.C., Benbasat, I.: Development of an instrument to measure the perceptions of adopting an information technology innovation. Inf. Syst. Res. 2(3), 192–222 (1991)
9. Venkatesh, V., Davis, F.D.: A theoretical extension of the technology acceptance model: four longitudinal field studies. Manag. Sci. 46(2), 186–204 (2000)
10. Venkatesh, V., Morris, M.G., Davis, G.B., Davis, F.D.: User acceptance of information technology: toward a unified view. MIS Q. 27(3), 425–478 (2003)
11. Venkatesh, V., Bala, H.: Technology acceptance model 3 and a research agenda on interventions. Decis. Sci. 39(2), 273–315 (2008)
12. Venkatesh, V., Thong, J.Y.L., Xu, X.: Consumer acceptance and use of information: extending the unified theory of acceptance and use of technology. MIS Q. 36(1), 157–178 (2012)
13. Oye, N.D., Iahad, A.N., Ab. Rahim, N.: A comparative study of acceptance and use of ICT among university academic staff of ADSU and LASU: Nigeria. Int. J. Sci. Technol. 1(1), 40–52 (2012)
14. Mac Callum, K., Jeffrey, L.: The influence of students' ICT skills and their adoption of mobile learning. Aust. J. Educ. Technol. 29(3), 303–314 (2013)

15. Shaper, L.K., Pervan, G.P.: ICT and OTs: a model of information and communication technology acceptance and utilizations by occupational therapist. Int. J. Med. Inform. **76**(1), 212–221 (2007)

16. Alrawashdeh, T., Muhairat, M., Alqatawnah, S.: Factors affecting acceptance of web-based training system: using extended UTAUT and structural equation modeling. Int. J. Comput. Sci. Eng. Inf. Technol. (IJCSEIT) **2**(2), 1–9 (2012)

17. King, W.R., He, J.: A meta-analysis of the technology acceptance model. Inf. Manag. **43**(6), 740–755 (2006)

18. Sedana, I.G.N., Wijaya, St.W.: UTAUT model for understanding learning management system. Internet Work. Internet J. **2**(2), 27–32 (2010)

19. Park, Y.: A pedagogical framework for mobile learning: categorizing educational applications of mobile technologies into four types. Int. Rev. Res. Open Distance Learn. **12**(2), 78–102 (2011)

20. Marchewka, J., Liu, C., Kostiwa, K.: An application of the UTAUT model for understanding student perceptions using course management software. Commun. IIMA **7**, 93–104 (2007)

21. Qin, C., Fan, B.: Factors that influence information sharing, collaboration, and coordination across administrative agencies at a Chinese University. Inf. Syst. e-Bus. Manag. **14**(3), 637–664 (2016)

22. Waheed M.: A Study of Teacher's Acceptance of e-Learning Technology: TAM as the Core Model. eGovshare (2009)

23. Pardamean, B., Susanto, M.: Assessing user acceptance toward blog technology using the UTAUT model. Int. J. Math. Comput. Simul. **6**(1), 203–212 (2012)

24. Liao, P.W., Yu, C., Yi, C.C.: Exploring effect factors of e-learning behavioral intention on cross-level analysis. In: Advanced Materials Research, pp. 204–210 (2011)

25. Arenas-Gaitá, J., Rondan-Cataluña, F., Ramirez-Correa, P.: Gender influence in perception and adoption of e-learning platforms. In: Proceedings of the 9th WSEAS International Conference on Data Networks, Communications, Computers, pp. 30–35 (2010)

26. Zanjani, F., Ramazani, F.: Investigation of E-learning acceptance in teaching english language based on TAM model. ARPN J. Syst. Softw. **2**(11), 130–135 (2012)

27. Islam, A.: Understanding Continued Usage Intention in e-Learning Context. AISeL (2011)

28. Sun, J.Y., Seung-hyun, H., Wenhao, H.: The roles of intrinsic motivators and extrinsic motivators in promoting e-learning in the workplace: a case from South Korea. Comput. Hum. Behav. **28**, 942–950 (2012)

29. Hsia, J.W., Tseng, A.H.: An enhanced technology acceptance model for eLearning systems in high-tech companies in Taiwan: analyzed by structural equation modeling. In: International Conference on Cyberworlds (2008)

30. Al-Gahtani, S.: Empirical investigation of e-learning acceptance and assimilation: a structural equation model. Appl. Comput. Inform. **12**(1), 27–50 (2016)

31. Ajzen, I.: Perceived behavioral control, self-efficacy, locus of control, and the theory of planned behavior. J. Appl. Soc. Psychol. **32**, 665–683 (2002)

32. Tuzel, S., Hobbs, R.: The use of social media and popular culture to advance cross-cultural understanding. Comunicar **51**, 231–242 (2017)

33. Sun, Y., Bhattacherjee, A., Ma, Q.: Extending technology usage to work settings: the role of perceived work compatibility in ERP implementation. Inf. Manag. **46**, 351–356 (2009)

34. Chin, W.: The partial least squares approach for structural equation modeling. In: Marcoulides, G.A. (ed.) Modern Methods for Business Research. Lawrence Erlbaum Associates, Mahwah (1998)

35. Tenenhaus, M., Vinzi, V.E., Chatelin, Y.M.Y., Lauro, C.: PLS path modeling. Comput. Stat. Data Anal. **48**, 159–205 (2005)

36. Kline, R.B.: Principles and Practice of Structural Equation Modeling, 2nd edn. The Guilford Press, New York (2005)
37. Sekaran, U.: Research Methods for Business: A Skill Building Approach, pp. 172–174. Wiley, Hoboken (2003)
38. Cronbach, L.J., Shavelson, R.J.: My current thoughts on coefficient alpha and successor procedures. Educ. Psychol. Meas. **64**(3), 391–418 (2004)
39. Thompson, B.: Understanding reliability and coefficient alpha. In: Thompson, B. (ed.) Score Reliability: Contemporary Thinking on Reliability Issues, pp. 3–23. Sage Publications Inc., Thousand Oaks (2003)
40. Zhang, P., Sun, H.: Affective quality and cognitive absorption: extending technology acceptance research. In: The Hawaii International Conference on System Sciences (2006)
41. Dasgupda, S., Haddad, M., Weiss, P., dan Bermudez, E.: User acceptance of case tools in system analysis and design: an empirical study. J. Inform Educ. Res. **9**(1), 51–78 (2007)
42. Chen, C., Wu, J., Yang, S.C.: Accelerating the use of weblogs as an alternative method to deliver case-based learning. Int. J. E-Learn. **7**(2), 331–349 (2008)
43. Sumak, B., Polancic, G., Hericko, M.: An empirical study of virtual learning environment adoption using UTAUT, pp. 17–22. IEEE (2010)
44. Sam, H.K., Othman, A.E.A., Nordin, Z.S.: Computer self-efficacy, computer anxiety, and attitudes toward the internet: a study among undergraduates in UNIMAS. Educ. Technol. Soc. **8**(4), 205–219 (2005)
45. Abbad, M., Morris, D., Nahlik, C.: Looking under the bonnet: factors affecting student adoption of e-learning systems in Jordan. Int. Rev. Res. Open Distance Learn. **10**, 1–23 (2009)
46. Lim, B.C., Kian, H.S., Kock T.W.: Acceptance of e-learning among distance learners: a Malaysian perspective. In: Proceedings ASCILITE Melbourne, pp. 541–551 (2008)
47. Kuciapski, M.: A model of mobile technologies acceptance for knowledge transfer by employees. J. Knowl. Manag. **21**(5), 1053–1076 (2017)

Language of Benefits as a Novel Tool for Improving Website Personalisation

Ossowska Katarzyna[(⊠)], Czaja Anna, and Sikorski Marcin

Department of Applied Business Informatics,
Faculty of Management and Economics, Gdansk University of Technology,
Narutowicza 11/12 street, 80-233 Gdansk, Poland
{Katarzyna.Ossowska, Anna.Czaja,
Marcin.Sikorski}@zie.pg.gda.pl

Abstract. A properly designed website allows the user to search for information faster, and more accurately. The information content of the website should be also adapted to the needs of the user. The purpose of this article is to present a novel Language of Benefits (LoB) approach to facilitate the use of websites for individual user groups. The LoB approach is an approach addressed to IT Analysts, to facilitate the process of web design, so that designed websites can better satisfy the many expected benefits of users themselves.

Keywords: Language of Benefits · Website development · Personalisation Requirements

1 Introduction

In the age of a Global Internet, the phenomenon of "information flood" is becoming more and more a reality. According to an IDC report, by 2025 the world will be creating 163 zettabytes of data per year [15]. Websites are filled with content that is tricky to navigate through, and it is becoming more and more difficult to find relevant information. Moreover, this phenomenon is often accompanied by other unwanted factors, such as:

- Dissemination of knowledge and information in an avalanche manner;
- Increasingly less visible boundaries between different types of knowledge and information;
- Globalisation of information and knowledge;
- Information chaos, which is very much related to the pace of change [12].

Organising information according to the usage of existing techniques, such as the User Experience [10], is no longer sufficient. Therefore, it seems correct to conduct further research into better adaptation of websites to the needs of their users. Nowadays, when creating websites, not only the client's requirements should be taken into account, but above all, users' expectations towards the website should as well. During the design phase, most importantly, the user should be provided with functions that will satisfy his/her expected benefits. However, there is still a lack of research on researching and obtaining the expected benefits.

S. Wrycza and J. Maślankowski (Eds.): SIGSAND/PLAIS 2018, LNBIP 333, pp. 86–102, 2018.
https://doi.org/10.1007/978-3-030-00060-8_7

This paper, directly affecting the issue of personalised websites, is an attempt to use **Language of Benefits – (LoB)** during the process of websites design and personalisation. Until now, the Language of Benefit was only a sales technique used in practice by sellers. There is no documentation, publication, or books about its use; there is also no scientific description of this language. LoB is a new, novel concept, first proposed by the authors for website design. The authors are still researching the possibilities of applying the LoB in various areas, especially in computer science. In this article, the purpose is to use this new approach as the answer to the ongoing need for better designer-user communication in interactive systems design.

In this paper the authors present the results of preliminary validation of this novel approach. The authors proposed a thesis that the LoB can, not only enhance the website design process, but also improve its quality by the better adaption of websites to the needs of the users.

This paper is divided into five parts. The first part is related to the process of creating websites and the methods of their personalisation. The second concerns the presentation of the genesis of the LoB and research, which has shown the necessity of its use in the process of creating websites. The third part shows how to use the LoB on an example website that the research was focused on. It shows how, with the help of a LoB, a website can be personalised and adapted to the expectations of its users. Due to the fact that the authors are just beginning research on the LoB, the last part contains further research that the authors would like to perform in the near future. It also focuses on the benefits of using LoB in relation to other techniques used to create and design websites.

2 Related Research

2.1 Method of Website Development

The website development lifecycle typically consists of five phases: planning and analysis, design, development, testing (validation), and installation and maintenance [3]. Depending on the adopted methodology, these phases might emerge one after the other, or as in the case of agile approach, adopt a more incremental manner.

The first stage is planning and analysis. This phase relies on gathering all the necessary requirements for the project, including understanding the client's goals for building the website, and assessing the competition. Through the use of stakeholder interviews, reviews of market research, user feedback, and other statistical data, a target audience is better defined. To comprehend the audience's needs and expectations, personas are constructed. Each persona is an archetypal representative of one of the customer segments. If carried out successfully, the number of personas shouldn't be greater than the three primary personas that represent the main target audience, and up to four secondary personas. Each persona is given a profile that consists of a photograph, a name, a description, demographic details, interests, area of expertise, and habits, etc. Knowing the profiles helps the development team understand users' motivations and therefore makes for better design decisions [2].

The second stage is design. During this phase a sitemap and wireframes are created. The sitemap shows the relationship between all key pages and defines the site's navigation. Wireframes are a sort of black and white blueprint for web pages. Without the distraction of colours or typeface choices it is easier to concentrate on planning the layout and the interaction between the page's elements [4].

According to Grid Theory, while making a wireframe one should take into consideration three main aspects:

- Technical constraints, like screen resolution or target browser version
- Business constraints, that is - the main goals of the project, for example, increasing visitor traffic, time spent on a site, conversion of site visitors to customers
- Content and editorial constraints, for example, the length of an article and the length of its headlines and summaries, images, interactive elements, etc. [9].

Only when the rules of navigation and layout are established is it time to make style tiles that are a visual reference to the site's design language. That includes: choosing a colour pallet that is appealing to the visitors, font types, coding techniques such as object size, shapes, and colour to mark information bits. The products of the design phase can also include sketches, HTML screen designs, prototypes, photo impressions, etc. The designers and project managers may then drive their decisions as to the final solutions that will be used in the next stage [4].

The third stage is development. The objective of this phase is to create a website solution that meets an approved design. During this phase HTML and CSS files are built, new content for the site is created, and the old one is refined [3].

Fourth stage is testing. The aim of this phase is to test the website, collect and analyse feedback, and fix detected issues. The first phase is development testing, in which small components of the system are tested, followed by system testing - when errors resulting from interactions between components are detected. The final step is acceptance testing, here the system is tested with real world data, and system requirements definitions are examined [16]. Depending on the selected methodology, the testing phase can start after, before, or during the development phase.

The fifth and last stage is deployment and maintenance. Once the product is formally approved by the client, a site can be launched on the production server. The aim of this phase is to keep the site alive and up to date with relevant content and a modern look and feel, make SEO adjustments, collect user feedback and react on it.

2.2 Methods of Website Personalisation

Websites personalisation allows for limiting the amount of insignificant information, thus making meaningful content more visible to the user. Personalisation is one of the stages in the Design Thinking method, the so-called empathisation stage [11]. In this stage personas are created. Personas are nothing more than a definite segment of users for whom a solution is being created. By creating a persona, this solution is more tailored to their needs, thus increasing the likelihood that users will benefit from it.

It is essential that the web site's design allows a high level of usability to the end user, as perceived usefulness has been proven to be an important factor for user's intention to return [18]. Careless design of information might increase users' cognitive load, resulting in slower performance and lower accuracy [19]. Relying on information gathered about the user, a website can be personalised according to his needs. This is achieved by filtering information, by adjusting interface, functionality and channel to the user's special needs. It is important to understand who the target group is (individuals or categories of individuals) and what model of personalisation should be used (implicit or explicit personalisation) [20]. While tailoring a website to individual preferences, one has to remember to give the user a sense of control. Interface that changes too unpredictably and rapidly leads to lower comfort levels, thus making the user visit the website less frequently [21].

3 Novel Approach for Website Personalisation

3.1 Genesis of the Novel Approach

The proposed model presented in the article is a result of the research that the authors carried out to check the usability and functionality of the website of Faculty of Management and Economics. This research was based on surveys carried out on users of the Faculty of Management and Economics.

A questionnaire, with both open-ended and closed questions, was chosen to carry out the survey, as it enables the collection of a vast sample of data, without missing the participant's personal convictions. To gather data the authors contacted 10 lecture staff and gave students a link to the survey Moodle Platform, thus eliminating the possibility of redundancy. The survey was conducted anonymously on a group of 83 students of Faculty of Management and Economics. Taking part in the research was voluntary, and the only incentive was the opportunity to improve the department's website.

The survey started with a few background questions, followed by preliminary validation of the participant's knowledge of the site, and their Internet surfing habits, and concluded with their suggestions for the site's improvement. In order to try to prevent respondents from not completing the survey, only closed questions were mandatory.

Among respondents, 50 participated in undergraduate studies and 33 in master's programs. The Department's website consists of seven main sections. Students were asked if they had visited any of the subpages of each section. If the answer was - "I don't use it (that particular section) at all", an additional question was asked -"why don't you visit it". The first question (Fig. 1) concerned visiting the FACULTY section. The most visited link proved to be - the Campus of the University site and the second one - the Structure of the Faculty. Other pages were of little interest to the sample group.

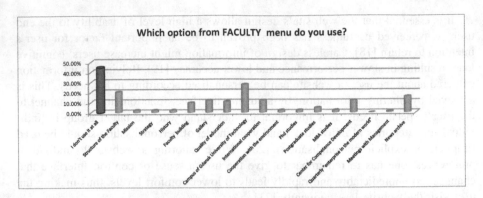

Fig. 1. Answers of 83 students to which option from faculty menu do they use

It turned out that over 30% of the students didn't feel the need to visit this section at all. When asked why they had not visited any of the FACULTY pages students answered that:

- I don't think the information there is relevant to me
- I don't feel the need
- Because I feel it is hard to find something there

The next question (Fig. 2) regarded the usage of subpages in the CANDIDATES section. Students were really only interested in one link called First and Secondary Studies.

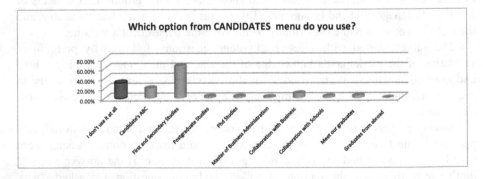

Fig. 2. Answers of 83 students to which option from candidates menu do they use

When asked why they had not visited any of the CANDIDATES pages students answered that:

- I don't think the information there is relevant to me
- I don't feel the need
- I'm not a candidate

- The site doesn't live up to my expectations
- The site is not intuitive

The most visited section proved to be the STUDENTS section (Fig. 4). Two thirds of the links put there got at least 20% of the users who visited them. To some surprise, even this area is not frequented by everyone.

- I don't feel the need
- I'm not interested in the information it contains

Unfortunately, according to this survey, students aren't significantly interested in pursuing an academic career, at least not at this point in their life. More than 83% of the group had never entered the SCIENTISTS section (Fig. 3).

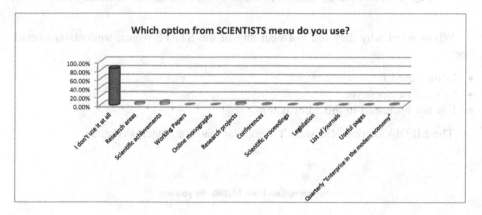

Fig. 3. Answers of 83 students to which option from scientists menu do they use

When asked why they had not visited any of the SCIENTISTS pages, students answered that:

- I don't need it
- I'm not a graduate
- I don't think the information there is relevant to me
- It's not addressed to me
- The title doesn't encourage me to

According to the survey, only the Employee list was interesting for the students to read on the STAFF section (Fig. 4).

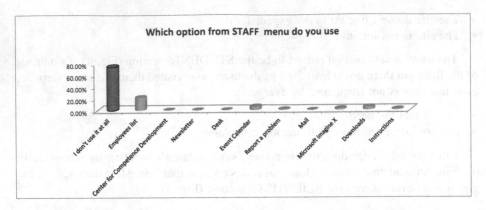

Fig. 4. Answers of 83 students to which option from staff menu do they use

When asked why they did not visit any of the STAFF pages, students answered that:

- I don't need it
- I'm not an employee
- I'm not interested in this information

The MEDIA section also wasn't a popular read for students (Fig. 5).

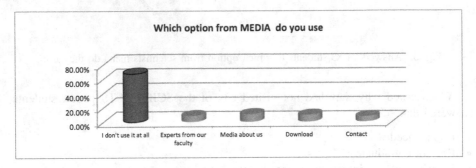

Fig. 5. Answers of 83 students to which option from media menu do they use

67% had never been to this section. When asked why they had not visited any of the MEDIA pages students answered that:

- I don't need it
- I'm not from media
- I'm not interested
- I didn't even know it existed, but I still think it's not relevant to me

Almost 69% of students had never felt the need to use the CONTACT section as opposed to 31% who did (Fig. 6).

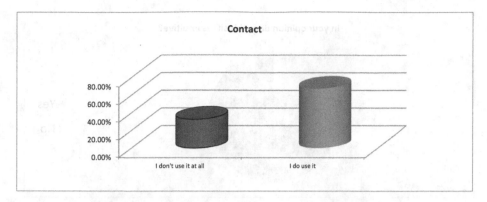

Fig. 6. Answers of 83 students about use contact menu

Most of the students had visited the website that week, or at least the same month the survey took place (Fig. 7).

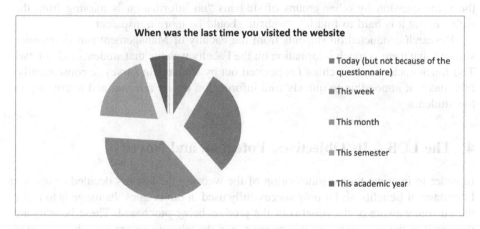

Fig. 7. Answers of 83 students about the last time they had visited the website

Even though students don't use a lot of department's page resources, 68% of them thought that the site is intuitive (Fig. 8). The results of this last question were so interesting that another survey has been conducted. This time it was a paper anonymous survey in which we asked 110 students what benefits they expect from the website. The most voted for was intuitiveness, fast responding, easy to navigate, and current.

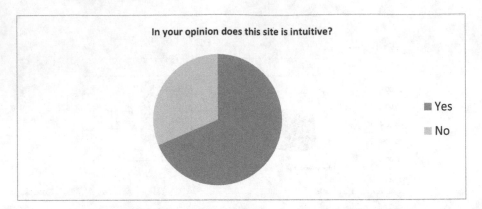

Fig. 8. Answers of 83 students about their opinion about site intuitiveness

Answering the question - "what information is missing from the department's website?" - students often indicated data that actually **was** on the website. The issue arises - how could they not have noticed it? Maybe a clue would be answers given to the same question by other groups of students "no information is missing from the website, but it is hard to find it", "website should be more transparent".

Research conducted on students from the Faculty of Management and Economics showed that there is much information on the Faculty website that students did not use. Too much content causes chaos (as pointed out by students in surveys); consequently, this makes it impossible to quickly find information that is relevant and interesting to the students.

4   The LOB – Its Objectives, Potential, and Novelty

In order to improve the personalisation of the website, the authors decided to use the Language of Benefits, so far only successfully used in direct sales. Its usage is to make the customer aware of the benefits of the product being purchased. These benefits are the result of the advantages of this product, and the advantages are directly connected with the features of this product. The product's features can be colour, shape, and size. The advantages: visibility, originality, and speed. The benefits: saving time, saving money, or increasing sales. The authors' idea is to use the Language of Benefits in the process of designing IT solutions, including personalisation of websites. The article presents the author's model of personalisation of websites. This model aims to adapt the website to the individual needs of each user (individual persona for each user). The authors assume that the personalisation of the website, according to only certain groups of users, may be insufficient, because each user may have other expected benefits, even though they belong to the same group of users.

The use of this approach may help increase the number of users making use of the website. This will only be possible by facilitating access to the information needed by the user. This approach is mainly aimed at IT analysts, and it facilitates the design of websites so they meet the expectations of potential users, as much as possible. Thanks to this approach, satisfying the expected benefits of users can be done in a much more individualised way.

4.1 Preliminary Validation of the LOB

The model proposed by the authors has been divided into three stages.

1. The first stage is the identification of the potential user. In this stage, as much information, as is possible, about who uses the website should be collected. This will allow for better adaptation to the user's needs and wants. For this purpose, on the website, the user is led through a decision tree, which consists of questions that are aimed at better identification. Questions should be constructed based on the specificity of the website, and its potential users. Questions should be constructed by those who have knowledge about the area associated with the website, and users who are interested in this area.
2. The second stage of the approach is to define the expected benefits. In order for the website to be encouraging to use, it must satisfy the benefits that users expect from it. If this is the case, they will be more willing to use it and return to it more readily. These benefits should be defined on the basis of website functionalities, and also defined by the users themselves. In this process, the analyst should consider which benefits each functionality can satisfy, but also give the user the ability to independently define the expected benefits.
3. The third and final stage is combining the expected benefits with specific functionalities of the site. This stage allows personalisation of the page based on the expected benefits. This allows the user to choose the benefits they expect from the site, and the use of the features that have been assigned to these benefits. Then the collection of benefits will be complete.

The authors decided to carry out an approach of preliminary validation, based on the website of the Faculty of Management and Economics. This fact is due to an excellent knowledge of the website, and the area of its operation, by the authors.

The preliminary validation was carried out in three stages:

1. Identification of users
2. Define users' expected benefits
3. Match the expected benefits with functionalities of the website

Personalisation of the website is done by selecting the option on the website (Fig. 9).

Fig. 9. Starting the page personalisation process (website prototype)

When the "Personalise the website" option is selected, the first stage is user identification. The authors distinguished the following groups: parents, students, administrative employees, the press, graduates, candidates, and others. This division was based on the authors' expert knowledge. Then, to better define the user of the website, questions were constructed. For example, in the case of the group "students", questions were related to mode, grade, and semester of study. These groups distinguish between different needs related to the degree of sophistication of the study process. Other functionalities will be used by part-time students (inter alia, dates of conventions), and full-time students, others by students of the first and last semester. In addition, the selection of a particular group results in the emergence of further identification questions related to a specific group (Fig. 10). For the purposes of this article, we assume that the user has selected the option "Student" during the first part of identification process.

It should be mentioned that each user's characteristic is related to the appearance of specific functions assigned to it. Assume that a user has chosen options: I Degree, Full-time studies and II Semester.

The next step is to choose the benefits. For this purpose, in the second stage, the benefits have to be defined first. To this end, the authors analysed the functionalities existing on the site and determined the benefits that these functionalities may satisfy. The question that the authors asked themselves during the analysis was: "What is the user's benefit from using this functionality?" In this way, the 10 benefits were defined as follows:

FACULTY OF MANAGEMENT AND ECONOMICS

PERSONALIZATION THE WEBSITE

Are you (choose the right one):

- o Student
- o Parent
- o Administration employee
- o Press
- o Graduate
- o Candidate
- o Other

FACULTY OF MANAGEMENT AND ECONOMICS

PERSONALIZATION THE WEBSITE

Your degree of study (choose the right one):

- o I
- o II
- o III

Your mode of study (choose the right one):

- o Full-time
- o Extramural

Your semester of study (choose the right one):

- o I
- o II
- o III
- o IV
- o V
- o VI
- o VII
- o VIII

Fig. 10. Identification process (website prototype – online survey)

1. Want only basic information
2. Scientific development
3. Want to know the latest information
4. Want to participate in the life of the university
5. Want to participate in non-university student life
6. Want to have a good job
7. Want to travel abroad as part of student exchange
8. Want to be as up to date with events as the faculty
9. Want to improve competences
10. Want to have better knowledge about the history of the faculty

Those benefits were placed in the next part of personalisation process (Fig. 11).

98   O. Katarzyna et al.

FACULTY OF MANAGEMENT AND ECONOMICS

PERSONALIZATION THE WEBSITE

Which benefits do you expect from using the website
(choose as many as you want):

o Want have only basic information
o Scientific development
o Want to know the latest information
o Want to participate in the life of the university
o Want to participate in non-university student life
o Want to have a good job
o Want to travel abroad as part of student exchange
o Be up to date with events as the faculty
o Want to improve competences
o Want to have better knowledge about the history of the faculty

Fig. 11. Choosing benefits by the user (website prototype – online survey)

After determining the benefits, the third stage is allocating the appropriate functionalities to the benefits. The functionalities were allocated to benefits by the expert, and they are already available on the website. The effects of this stage are presented in the Table 1.

Table 1. Matched benefits and functionalities

Benefit	Functionalities
1. Only basic information	Classes, news, tuition fees, diplomas, internships, contact, dean's office
2. Scientific development	Research projects, conferences, scientific proceedings
3. Know the latest information	News, Blogs, information for students, Facebook
4. Participation in the life of the university	IPMA student, scientific clubs, faculty students' council, MOST program
5. Participation in non-university student life	Student life
6. Have a good job	Job offers for students, internships
7. Travel abroad as part of student exchange	International cooperation
8. Be up to date with events as the faculty	Calendar
9. Improving competences	Centre for competence development, projects office
10. Have better knowledge about the history of the faculty	About the department

On the basis of the functionalities assigned to the benefit, the user, by choosing a given benefit, receives a set of functions assigned to this benefit. For the purposes of this article, we assume that the user (student) chose the following benefits: Have a good job, Participate in non-university student life. Consequently, the user's personalized page will look as shown in the Fig. 12:

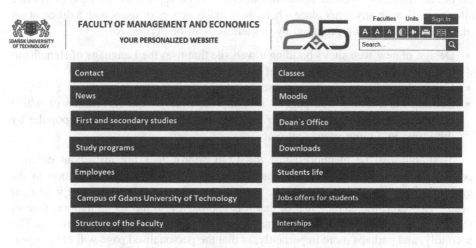

Fig. 12. Personalised page (website prototype)

Of course, the user cannot be deprived of access to other functionalities of the site, which is why the authors propose that the personalised page should be only a subpage (addition to the page) that can be selected at any time by the user.

5   Discussion and Further Research

The approach outlined in this article allows for the analysis of the functional requirements of a website from the user's point of view, using the Language of Benefits.

This paper presents only preliminary validation of an approach that uses the Language of Benefits to personalise websites. Initial validation was aimed at obtaining feedback from users and website designers. At this point, it should be emphasised that the approach was met with interest from the website administrator. Thanks to this, further research will be possible, assuming, for example, creation of a test page that will be presented to a wider range of users (parents, press, and employees). The website administrators have also indicated that the main problem is the information overload on the website, and this approach can be solved in a very accessible way. Also, a study conducted on students of the Faculty of Management and Economics showed that there is a great deal of information on the faculty's website that students do not use. Too much content causes chaos (as pointed out by students in surveys), and consequently, it makes it impossible to quickly find information that is relevant and interesting to students. This group of users also indicated that there is a need to make changes on the website, so that it becomes more legible, and a way to acquire information faster and more intuitively.

The LOB concept needs further testing and validation. This approach is now based on public university website research, but can also be applied to other public service sites as well, as in any other place where there exists groups of users with a different set of needs. The approach proposed by the authors can be applied to:

- Design of new web sites - building a web site that uses the Language of Benefits to define functionality and location of the site
- Verify the website to better suit user needs
- Redesign of existing web pages - this approach allows the verification of which elements are used by different user groups, and promotion of the most popular by the group to a more prominent place

It should also be mentioned that the LOB approach is not just about defining personas. Thanks to this approach, each user, regardless of his/her affiliation to the group, interest, or goal in visiting the website, will be able to personalise it to their needs. However, the authors realize that this approach also has its limitations. One of them may be the users' lack of willingness to go through the identification process. The secondly, and perhaps more importantly, is that the personalised page will only appear on the IP address on which the user accessed it.

Later research on LoB will also cover other areas, for example, IT project management. Due to the fact that LoB can improve communication between the client and the project team, the authors see the possibility of using LoB especially in agile projects, where communication is ubiquitous (Fig. 13).

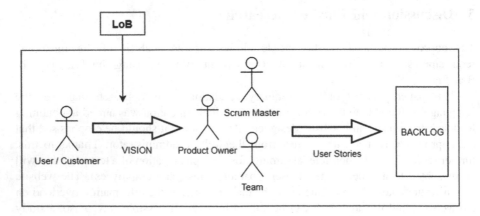

Fig. 13. Possible usage of LoB in agile projects

There is also a need for a formal LOB specification (graphic notation, syntax, diagrams, classes of benefits, communication channels, user profiling) that authors want to develop, after the completion of subsequent (more complex) case studies. The LoB should have defined its components - benefit dictionaries, process owners, such as User, Customer, and Project Team as a Solution Vendor. Such formal specifications can also be used in other areas and languages where it will be possible to use the LoB.

In regard to other languages, it is worth noting that the LoB can also be a complement to the UML (Unified Modelling Language) language associated with modelling, including system requirements, which are directly related to the benefits that they satisfy.

6   Conclusions

Personalisation of websites, in an era of overload of various types of information, is a problem of great importance and demands a proposal for new approaches and methodologies.

This article presents the initial validation of the novel approach, which uses the Language of Benefits (LoB) to better design websites. In the first part, the authors briefly presented other methods of personalisation of websites, as well as the process of their creation. The second part is the origin of the approach - this is very important from the point of view of the authors' motives - to the authors' interest in the problem of personalising websites. The third part presents the LoB approach, using the existing website of the Faculty of Management and Economics and its prototypes as an example, after applying the LoB approach. The authors, while aware of the limitations and the need for further research on the approach, in the fourth part presented the possibilities of further research in the future.

References

1. Bailey, R.W., Wolfson, C.A., Nall, J., Koyani, S.: Performance-based usability testing: metrics that have the greatest impact for improving a system's usability. In: Kurosu, M. (ed.) HCD 2009. LNCS, vol. 5619, pp. 3–12. Springer, Heidelberg (2009). https://doi.org/10. 1007/978-3-642-02806-9_1
2. Lidwell, W., Holden, K., Butler, J.: Universal Principles of Design, Revised and Updated Edition, 2nd edn. Rockport Publishers, Beverly (2010)
3. Darlington, K.: Effective Website Development: Tools and Techniques. Pearson Education, Toronto (2005)
4. Lopuck, L.: Web Design for Dummies. Wiley, Hoboken (2012)
5. Smashing Magazine: Professional Web Design: The Best of Smashing Magazine (2011)
6. Johnson, J.: Designing with the Mind in Mind: Simple Guide to Understanding User Interface Design Guidelines. Elsevier, Boston (2013)
7. Walter, A.: Designing for Emotion (2011)
8. Watts, D.: Is Justin Timberlake a Product of Cumulative Advantage? (2007)
9. Vinh, K.: Ordering Disorder: Grid Principles for Web Design (2010)

10. Dix, A., Finlay, J., Abowd, G., Beale, R.: Human-Computer Interaction. Prentice Hall, Upper Saddle River (2004)
11. Lowdermilk, T.: User-Centered Design: A Developer's Guide to Building User-Friendly Applications. O'Reilly, Beijing (2013)
12. Gawrysiuk, P.: Cyfrowa rewolucja. Rozwój cywilizacji informacyjnej (2008)
13. Ossowska, K., Szewc, L., Weichbroth, P., Garnik, I., Sikorski, M.: Exploring an ontological approach for user requirements elicitation in the design of online virtual agents. In: Wrycza, S. (ed.) SIGSAND/PLAIS 2016. LNBIP, vol. 264, pp. 40–55. Springer, Cham (2016). https://doi.org/10.1007/978-3-319-46642-2_3
14. http://www.nytimes.com/2007/04/15/magazine/15wwlnidealab.t.html
15. https://www.forbes.com/sites/andrewcave/2017/04/13/what-will-we-do-when-the-worlds-data-hits-163-zettabytes-in-2025/#cfcd8f3349ab
16. Sommerville, I.: Software Engineering, 9th edn. Person Education Limited, Harlow (2011)
17. https://www.google.com/analytics/
18. Koufaris, M.: Applying the technology acceptance model and flow theory to online consumer behavior. Inf. Syst. Res. 13(2), 205–223 (2002)
19. Streveler, D.J., Wasserman, A.I.: Quantitative measures of the spatial properties of screen designs. In: Shackel, B. (ed.) Proceedings of Interact 1984 Conference on Human-Computer Interaction. Elsevier, Amsterdam (1984)
20. Marshall, H.F., Poole, S.: What is personalization? Perspectives on the design and implementation of personalization in information systems. J. Organ. Comput. Electron. Commer. 16(3&4), 179–202 (2006)
21. Sánchez-Franco, M.J., Rodríguez-Bobada Rey, J.: Personal factors affecting users' web session lengths. Internet Res. 14(1), 62–80 (2004)
22. PMBOK Guide, 6th edn, Newtown Square (2017)

Internet of Things and Big Data

IoT and Its Role in Developing Smart Cities

Catalin Vrabie(⊠)

Faculty of Public Administration,
National University of Political Studies and Public Administration,
Bucharest, Romania
vrabie.catalin@gmail.com

Abstract. A main characteristic of smart cities is the use of information and communications technology in all aspects of city life. In this regard, Internet of Things (IoT) is a core element in the process of developing communities "ruled" by an improved communication, better understanding and wait times decrease. This paper aims to present the ways in which IoT networks and services can contribute to develop smart cities, giving as example various cities that have implemented this concept. The methodology used to carry out this research is both bibliographic – opting here to study the work of specialists in the field, authors from Romania and abroad, and empirical – formed by a case study on various smart cities around the world that use IoT. This type of smart cities is starting to transform all public institutions, changing their culture, from one control-based to one performance-centered. IoT is starting to play an important role in smart cities' evolution and it brings an improvement in the government-citizens relationship. We have identified that although technology is a central element, there should also be considered the capability and willingness of citizens and public institutions to collaborate in order to implement the best solutions for the communities.

Keywords: Internet of Things · Urban development · Technology

1 Introduction

The Internet of Things concept (known in the literature as IoT) is not as new as one might think. It first appeared in 1999 when Kevin Ashton – the British who created the RFID (Radio Frequency Identification) systems standards, used it to describe a system in which the Internet connects to the physical world through sensors [2], these having the role of collecting data for sending them over networks to servers. Since back then he described how the devices connected to the Internet will change our lives, which nowadays is already far from being science-fiction. We see everywhere around us either cars connected to the Internet (via GPS terminals installed on board), industrial or agricultural equipment remotely coordinated through the Internet, drones, even refrigerators and washing machines (the smart mobile phones, present in everyone's pocket, are the best proof of the development of this IT industry's segment).

Today, the total number of connected equipment reached 20.35 billion, with the prospect of reaching 75.44 billion in 2025 [18].

© Springer Nature Switzerland AG 2018
S. Wrycza and J. Maślankowski (Eds.): SIGSAND/PLAIS 2018, LNBIP 333, pp. 105–113, 2018.
https://doi.org/10.1007/978-3-030-00060-8_8

The main components of an IoT system are the following [21]:

- **Data collection equipment** – some examples here would be: sensors, mobile phones, etc.;
- **Communication networks** with the role of connecting the equipment mentioned above – such as Wi-Fi, 4G, Bluetooth etc.;
- **Servers and other computational systems** that use these data – such as: storage, analysis devices or dedicated software applications.

When all three of these components are found in the same system with the role to deliver services (and sometimes products), then we can really talk about added value created with the aim of developing citizens, the public and the private environment [25]. A short example would be the smart devices that monitor the evolution/involution of a chronic disease in a patient by transmitting real-time data to doctors who may intervene if the situation requires so.

IoT applications and systems are organically developed – based on needs, but the impact they have on us depends on the degree of acceptance of new technologies by citizens, the public and the private sector [22].

The greatest risks that can arise from the extensive use of IoT come from the data security and cyber-attacks area. However, the laws of the economy must be understood, namely that the most trustworthy products and services will continue to be procured by the beneficiaries – demand and supply are strongly connected. The Statistic Portal tells us that the IoT market has exceeded a trillion dollar at the end of 2017, forecasting an evolution of up to 1.7 trillion dollar at the end of 2019 [18].

2   Cities with Senses

More and more cities in the world are experiencing the new dimension of sensor networks. Many are involved in pilot projects with the purpose of monitoring various activities in urban life, such as the level of noise or air pollution, parking management, health monitoring applications for persons suffering from chronic illnesses etc. **Thingful** is a search engine within this new dimension of the digital world. It contains indexes with the geographical positioning of all the fixed equipment connected in the world – a simple typing of a city's name can indicate on the map where different sensors are placed and what function they fulfil [20].

Thingful's goal is not just to provide a map of existing public or private equipments, but also to provide developers with solutions for smart cities to use these devices – of course, with the consent of the owners [20] (Fig. 1).

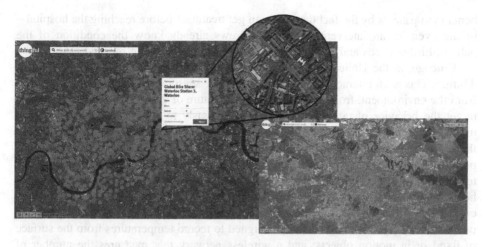

Fig. 1. World of IoT in London, UK (left) and Bucharest, Romania (right) [20]

London has developed with six partners, including Future Cities Catapult and Intel Collaborative Research Institute, the project Sensing London. Five living labs were built around the metropolis to collect data (obviously through sensors) about humidity, air quality, traffic and pedestrian activity. Subsequent analyzes directly help enrich the knowledge of how British capital residents use the infrastructure. At the same time, indirectly, these data are used as inputs in the health, environment and life comfort systems due to the statistical analyzes that can be carried out and thus the impact that a particular phenomenon can have in the area of interest researched can be predicted. From this point to developing new solutions (such as an application that would help asthmatic patients to travel through the city) or to developing new business models that allow the expansion of green spaces without major financial investment or even the justification for the development of new technological infrastructures is just one step [10].

The Christchurch city of New Zealand has developed, through a nonprofit organization, a similarly project called **Sensing City Trust**. The actors involved want to better understand how data gathered through sensors can help mayors to develop better public policies. After a devastating earthquake in 2011, a network of digital sensors was developed and installed as part of the city's physical infrastructure in order to gather information on air quality. In addition, 150 people registered in the public health system were recruited as chronic respiratory patients who were given a "smartinhaler" which records where and when they are using medication to relieve symptoms. The data is then automatically transmitted via the smart phones the individuals owns, to a secured database, overlapping those that come from the sensors we mentioned and which were collected shortly before, and thus offered to decision makers for them to be able to develop the most effective public health policies. Supplementary to the initial purpose of the project, the information produced by the analyzes help doctors to improve their understanding of chronic lung diseases, thus managing to bring real benefits to patients by the fact that they can get treatment before reaching the hospital – in the event of an intervention, medical crews already know the condition of the patient, his/her needs and implicitly their response time is being reduced [15].

Chicago, in the United States, has developed a matrix of equipment – **Array of Things.** This is an interactive network of modular sensors that collects real-time data from the environment, from the physical infrastructure of the city, as well as those that target the behavior of citizens. The goal is obviously to better understand the local urban environment and the impact it has on the lives of individuals living and working there – the most important elements for analyzes being those related to climate, air pollution and noise pollution. The data thus collected are open, meaning that are open to free use by residents, software developers, scientists or decision-makers. Citizens' behavior is detected through three different types of sensors: sound sensors, which collect data from the surrounding environment; infrared cameras oriented to car or pedestrian traffic areas and which are designed to record temperatures from the surface of fixed or in motion objects; and a wireless network that measures the number of nearby Bluetooth and Wi-Fi devices – it acts as a proxy for pedestrians in the area. Although questions can be raised that would concern the privacy area of citizens, the project guarantees that no personal or identification data are collected [1] (Fig. 2).

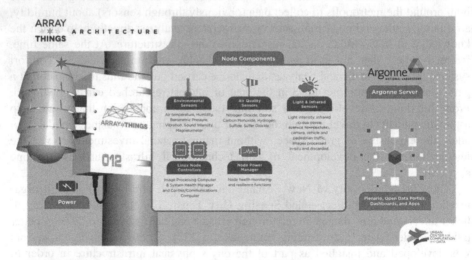

Fig. 2. Architecture of the array of things system [1]

In Sibiu, Romania, was developed, thanks to the collaboration of "Lucian Blaga" University of Sibiu with the University College of Southeast Norway, Norway, the project **A Mobile Platform for Environmental Monitoring** with the aim of producing an environmental map that provides all actors in the city's perimeter with information on air quality and noise pollution. The Faculty of Engineering within "Lucian Blaga" University has developed hardware modules that can be placed on cars and which are meant to collect traffic data both when the car is in motion or in the parking areas, when the car is parked. The collected data is transmitted via a GSM module combined with a

GPS module implemented on the equipment to a server that has the role of storing them and providing them for analysis to the actors involved. Two prototypes of sensors were realized, the last of them (and the most advanced) being able to collect both data such as the CO2, NOx level as well as the amount of suspended solid particles. The project is still in the pilot phase, with only 16 cars equipped with such modules in the city, being completely functional, a number of approximately 100 units will be produced to be mounted on vehicles [4] (Fig. 3).

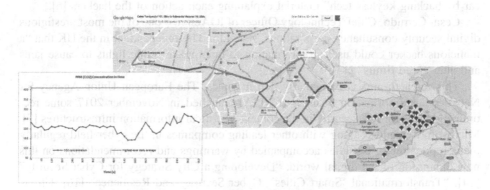

Fig. 3. Data collected through mobile traffic platforms in Sibiu [8]

3   The Risks of a Connected City

3.1   Cyber Security

As our cities are becoming more and more saturated with sensors, they are becoming smarter and smarter [5]. However, we must also take into account citizens' degree of tolerance for the invasion of data collection equipment – as the number of equipment increases, the citizens feel more supervised [23].

The most common questions here are: (1) "Who produces and controls the equipment?", (2) "What do they measure?", and (3) "Who has access to the data?". All these questions are important and answers to them must be available to every citizen in a language that is as easy to understand as possible so that there is no confusion.

Other questions such as those related to the purpose of collecting data, the changes that will follow from these operations and the benefits of citizens, the public and the private sector are also important. Data storage management mechanisms (often software) are also commonly found in studies about IoT.

Many cities consider elements of security (obviously not only digital) and intimacy as key to sustainable and harmonious development. The level of trust and acceptance of the new by citizens is crucial in developing smart solutions. However, there is little written information on how citizens see these things.

Dan Gârlaşu, from Oracle Romania, warned IT users that in the future smart cities may be more vulnerable to hackers than smart computers and smartphones are today [11].

With billions of interconnected devices all over the world, cyber security challenges are increasingly addressing also the IoT dimension of the digital world. Often the media poses on the front page of the newspapers titles that refer to hacking actions of different types of equipment. In the summer of 2015, the car producer Fiat recalled 1.4 million vehicles for software updates due to the risks of the machine safety being affected [3]. At the end of 2017, a clip posted on Youtube featured two hackers who stole a luxurious car by remotely cloning the door opening device and starting the vehicle [24]. Shortly after the event, CNN tech has produced the "Watch thieves steal car by hacking keyless tech" material explaining each action of the hackers [6].

Cesar Cerrudo, Chief Technology Officer of IOActive – one of the most prestigious digital security consultancy corporations, stated for The Independent in the UK that "a malicious hacker could use the information to manipulate traffic lights to cause jams and alter speed limits" [13].

This research area is particularly rich in topics. The European Union Agency for Network and Information Security (ENISA) launched in November 2017 some recommendations on IoT security in the context of critical information infrastructures [7]. Microsoft, Symantec, along with other leading companies in the cyber field regularly make reports on case studies accompanied by warnings and recommendations on this new dimension of the digital world: "Developing a City Strategy for Cyber Security" [14], "Transformational 'Smart Cities': Cyber Security and Resilience" [16]. Unfortunately, however, there is little information on how these recommendations are embedded in smart solutions implemented at city level.

EU is acting by renewing its cybersecurity strategy, firstly by transforming ENISA (European Union Agency for Network and Information Security) into a EU cybersecurity agency 'able to prevent and respond to cyber-attacks in a more coordinated way'. By that the agency will be able to conduct pan European cybersecurity exercises and will ensure a better sharing of intelligence [29].

Secondly, EU is promoting 'the creation of a true Single Cybersecurity market with an EU-wide framework for cybersecurity certification'. This initiative is crucial due to the rapid development of the IoT world [29].

Cyber security efforts tend to be focused on the role of local leaders in the development of smart cities and the IT&C embedded systems, although it is known that the development of such cities is much more complex, involving many partners in this process and as many technologies. For example, The U.S. Chamber of Commerce is 'optimistic about the future of the IoT, which continues the decades-long trend of connecting networks of objects through the Internet. The IoT will significantly affect many aspects of the economy, and the Chamber wants to constructively shape the breadth and nature of its eventual impact' [28].

3.2   Temporary Inoperability

IoT enthusiasm is often tempered by the connectivity problems that the equipment are faced with. The wireless ecosystem, though easy to understand, is hard to imagine. Due to the very large number of IoT uses, we cannot find a single standard – both in wireless technologies and in electricity consumption [17]. These two seemingly minor problems can cause major effects in the good functioning of an IoT system.

The technology of a smart city could be taken by surprise by the technological advances – new equipment is being developed, with new standards long before the old and already in motion ones are depreciated [26]. Hence, many connectivity problems can arise between the equipment placed in the wireless ecosystem of IoT. For example, we can imagine a smart city in which automated cars (without a driver) navigate by themselves on the city's streets. What happens when they pass through an area in which the sensors of the traffic lights are no longer compatible with theirs? Another question that arises is what happens when, due to network noise, communication between the vehicle and the traffic light system is slow or temporarily interrupted? Obviously, these questions must first find an unequivocal answer in order to be able to talk about a successful implementation of such a system [12]. In Fig. 4 we can see the complexity of such a system and, practically, due to the large number of devices that need to communicate in a very short time, the risks associated to a small data flow disruption.

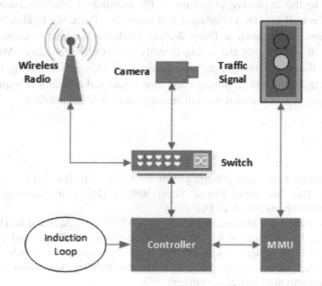

Fig. 4. A typical traffic intersection [12]

All Internet users experienced situations where web pages were loading slower or when mobile calls were disrupted apparently for no reason. These situations can create frustration, but humans understand and know they can appear. But when we talk about electronic equipment, they cannot understand, and the effects of their misunderstanding can produce less pleasant effects for citizens or the environment [27].

If, in the case of the cyber security risks previously presented, the pressure was on the managers of a smart city, in the case of interoperability, the pressure tends to be put on the research environment, especially in the technological and academic areas. Only these can find viable solutions to such problems.

4 Conclusions

The dimension of IoT is not just a goal to be achieved – often mayors, hearing the concept but not understanding it in its depth, want to invest in IoT sensors and equipment for their cities – it is a remarkable symbiosis between society and technology. Many of those technologies that once represented the top ones are today viewed as part of everyone's life.

The parallel between IT and Internet innovations has led to a series of changes in the world economy such as the growth of the sector of products and services dedicated to informational economy. As Thomas Lauren Friedman (a New York Times journalist) said in his book "The world is flat: A brief history of the twenty-first century", written in 2005, "the Internet has flattened the world, IT has first provoked and then increased the pace with which these changes have occurred, providing a platform for development" [9].

Of all the challenges of the electronic world, IoT is the newest and probably the biggest – due to the explosive evolution of the number of Internet-connected equipment. It must be well known, understood and managed. There is a hidden component in people's Internet, also known as Deep Web or Dark Web, where unknown operations are made and of which only the actors directly involved have a clue. Many of these operations are illegal. Why wouldn't the Internet of Things risk having its own dark side? To minimize this risk, a proper education of all stakeholders is required, so that the responsibility for successful system management will be implicit.

References

1. Array of Things. https://arrayofthings.github.io/. Accessed 10 Dec 2017
2. Asthon, K.: That 'Internet of Things' Thing. RFID J. (2009). http://www.rfidjournal.com/articles/view?4986. Accessed 10 Dec 2017
3. BBC News. http://www.bbc.com/news/technology-33650491. Accessed 10 Dec 2017
4. Berntzen, L., Johannessen, M.R., Florea, A.: Smart cities: challenges and a sensor-based solution, a research design for sensor-based smart city projects. Int. J. Adv. Intell. Syst. **9** (3&4) (2016). International Academy, Research and Industry Association (IARIA) http://www.iariajournals.org/intelligent_systems
5. Business Insider: How smart cities & IoT will change our communities (2016). http://www.businessinsider.com/internet-of-things-smart-cities-2016-10. Accessed 10 Dec 2017
6. CNN Tech: Watch thieves steal car by hacking keyless tech (2017). http://money.cnn.com/video/technology/2017/11/28/relay-box-car-theft.cnnmoney/index.html. Accessed 10 Dec 2017
7. ENISA: Baseline Security Recommendations for IoT (2017). https://www.enisa.europa.eu/publications/baseline-security-recommendations-for-iot. Accessed 10 Dec 2017
8. Florea, A., Bertntzen, L.: Green IT solutions for smart city sustainability. Paper presented at the Smart Cities Conference, 5th edn, 8 December 2017. SNSPA, Bucharest (2017)
9. Friedman, T.L.: The World is Flat: A Brief History of the Twenty-First Century. Farrar, Straus and Giroux, New York (2005)
10. Future Cities Catapult: Sensing London. http://futurecities.catapult.org.uk/project/sensing-london/. Accessed 10 Dec 2017

11. Gărlaşu, D.: Cyber security update on threats and trends. Paper presented at the Smart Cities Conference, 4th edn, December 2016. SNSPA, Bucharest (2016). http://administratiepublica. eu/smartcitiesconference/2016/program.htm
12. Ghena, B., Beyer, W., Hillaker, A., Pevarnek, J., Halderman, J.A.: Green lights forever: analyzing the security of traffic infrastructure. In: Proceedings of the 8th USENIX Workshop on Offensive Technologies (WOOT 2014), August 2014 (2014)
13. Independent, Vulnerabilities in traffic light sensors could lead to crashes, researcher claims (2014). http://www.independent.co.uk/life-style/gadgets-and-tech/news/traffic-light-hack-could-lead-to-road-chaos-claims-expert-9309936.html. Accessed 10 Dec 2017
14. Microsoft Corporation: Developing a City Strategy for Cyber Security. A Seven-Step Guide for Local Governments (2014)
15. Sensing City. http://sensingcity.org/. Accessed 10 Nov 2017
16. Symantec Official Blog: Transformational 'Smart Cities': Cyber Security and Resilience (2013). https://www.symantec.com/connect/blogs/transformational-smart-cities-cyber-secu rity-and-resilience. Accessed 10 Dec 2017
17. Texas Instruments: Wireless connectivity for the, Internet of Things: One size does not fit all (2017)
18. The Statistic Portal: Internet of Things (IoT) connected devices installed base worldwide from 2015 to 2025 (in billions) (2017). https://www.statista.com/statistics/471264/iot-number-of-connected-devices-worldwide/. Accessed 10 Dec 2017
19. The Statistic Portal: Size of the global Internet of Things (IoT) market from 2009 to 2019 (in billion U.S. dollars) (2017). https://www.statista.com/statistics/485136/global-internet-of-things-market-size/. Accessed 10 Dec 2017
20. Thingful. http://www.thingful.net/. Accessed 10 Dec 2017
21. UK RS Online. https://uk.rs-online.com/web/generalDisplay.html?id=i/iot-internet-of-things. Accessed 10 Nov 2017
22. Vrabie, C.: Elements of E-Government. Pro Universitaria Publishing House, Bucharest (2016)
23. Vrabie, C.: Your Freedom Starts Where My Privacy Ends, Smart cities. Pro Universitaria Publishing House, Bucharest (2017)
24. YouTube. https://www.youtube.com/watch?v=bR8RrmEizVg. Accessed 11 Dec 2017
25. Verovsek, S., Juvancic, M., Zupancic, T.: Data-driven support for smart renewal of urban neighbourhoods. Smart Cities Reg. Dev. (SCRD) J. [S.l.] 2(2), 91–100 (2018). ISSN 2537-3803. http://scrd.eu/index.php/scrd/article/view/41
26. Nemţanu, F., Pînzaru, F.: Smart City management based on IoT. Smart Cities Reg. Dev. (SCRD) J. [S.l.] 1(1), 91–97 (2017). ISSN 2537-3803. http://scrd.eu/index.php/scrd/article/view/12
27. Barsi, B.: Beyond indicators, new methods in Smart city assessment. Smart Cities Reg. Dev. (SCRD) J. [S.l.] 2(1), 87–99 (2018). ISSN 2537-3803. http://scrd.eu/index.php/scrd/article/view/31
28. Eggers, M.J.: IoT Cyber Policy, U.S. Chamber of Commerce, NIST IoT Cybersecurity Colloquium. https://www.nist.gov/
29. Bieńkowska, E.: Charter of trust for a secure Digital World. In: Keynote Speech at Munich Security Conference, 16 February 2018. ec.europa.eu

Proposing a Behavior-Based IDS Model
for IoT Environment

Fadi Abusafat[(⊠)] , Tiago Pereira , and Henrique Santos

Algorithm Center, University of Minho, 4710-570 Guimarães, Portugal
Faabusafat1987@hotmail.com

Abstract. In the 21th century, Internet of Things (IoT) is considered an important topic into Information Communication and Technology due to massive revenue and uses in several fields. It has several communication protocols, multiple technologies implemented into its objects, also interactions between IoT applications and users depends on Wireless Sensor Networks technology (WSN). Therefore, improving IoT depends on developing the WSN. WSN is based on spread nodes that are connected with several sensors in order to collect and share information to provide its functionalities. However, there are several threats, vulnerabilities and attacks that affect functionality of WSN. These kind of attacks is considered sophisticated and very difficult to discover and detect. Therefore, there is the need to measure WSN behavior in order to indicate abnormal behavior and classify it as an attack. In this paper, we proposed a model to measure WSN behavior based on attacks effects. Also, this model consists of metrics in order to define a security architecture for Smart City Comprising Intrusion Detection mechanism to detect major attacks.

Keywords: WSN Metrics · Smart City · IoT Security · Intrusion Detection
WSN Security Treatment

1 Introduction

Internet of Things (IoT) is considered one of the most important topic into Information Communication and Technology (ICT) [21]. According to several studies, revenue of adapting it will be around 14.4\$ trillion between 2013 and 2022. This revenue reflects it used in several domains such as Transport, Smart grid, Smart City and others [35].

IoT depends on Wireless Sensor Network (WSN) which consists of several nodes that are able to sense, transmit and sharing data [14]. Also, it involves into several applications such as industrial, security, healthcare and others [34]. The main characteristics of WSN are constraint power consumption, deal with failure node, mobility, heterogeneous and scalability. These characteristics are introduced several security requirements such as Data Confidentiality, Integrity, Authentication, Availability, among others [26]. Besides that, it introduced several vulnerabilities such as sophisticated attacks. There are several classifications of attacks such as attacks based on devices, Active, Passive and Location. The effects of these attacks are between interference and stop services [10].

© Springer Nature Switzerland AG 2018
S. Wrycza and J. Maślankowski (Eds.): SIGSAND/PLAIS 2018, LNBIP 333, pp. 114–134, 2018.
https://doi.org/10.1007/978-3-030-00060-8_9

Intrusion Detection mechanism (IDs) is considered an adequate tool or software for protecting networks from attacks. It operates through monitoring host or network and alarm administrator when security violation is detected [35]. However, IDs still in the development stage and there are several issue needs to improve such high accuracy and low false alarm. In the WSN context with many communication protocols such as RF, TCP, IP, SNMP, FTTP, HTTP and many IoT devices from street sensor to smart grid, this kind of issues are potentiated.

In this paper, we identify measurement metrics for WSN based on the attack behavior in order to improve functionality of IDs. This paper is organized as follows: Sect. 2: overview of WSN, Sect. 3: Security into WSN, Sect. 4: WSN metrics, Sect. 6: Future works.

2 Overview of Wireless Sensor Network (WSN)

2.1 Definition and Architecture of WSN

WSN is defined as a network that communicates through wireless technology. This technology depends on spread nodes that is connected with several sensors in order to collect and share information between users and servers. Therefore, it is considered as a heterogeneous system due to it consists of several sensors and objects [13]. In IoT environment, nodes communicate independently in order to transmit data that's collected by the sensor to the gateway or base station. The main role of gateway is to transmit data from the wireless network to the main centre through network [13]. The main components of WSN are clarified in Fig. 1 and these are: sensor, processing, memory, power and wireless transceiver unit. The main roles of WSN are routing, broadcast, multicast, forward and maintains [23].

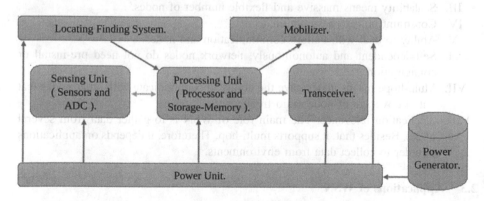

Fig. 1. Components of WSN [23].

However, the components of the architecture of WSN are sensors, network and security administrator, aggregation point and base station which includes access point and gateway Fig. 2. Gateway, is responsible to forward packets from access point to base station [23].

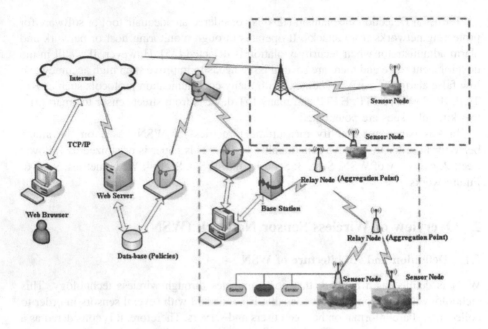

Fig. 2. The architecture of WSN [23].

2.2   Characteristics of WSN

There are several characteristics of WSN [9, 18, 22] such as:

    I. Dynamic topology.
   II. Heterogeneous, several types of nodes and devices.
  III. Scalability means massive and flexible number of nodes.
  IV. Constraint power consumption.
   V. Ability to deal with Failure communication and nodes.
  VI. Self-dependent and autonomously, network nodes do not need pre-install or configuration.
 VII. Multi-hope means nodes have the ability to communicate with other nodes that out of coverage of node radio frequency.
VIII. Application relevance. The main role of WSN is to gather data from several nodes. Besides that, it supports multi-hop. Therefore, it depends on applications in order to collect data from environments.

2.3   Applications of WSN

Applications into WSN are wide. Generally speaking, whole of them are categorized into two fields, monitoring and Tracking. First one used for measurements such as pressures, temperatures, humidity…etc. The second one for pursuit movement of objectives such as distance, speed, arriving time…etc. However, there are several divers for WSN application based onto operating environment or areas such as military, public, health, education, environments …etc. Fig. 3 [34].

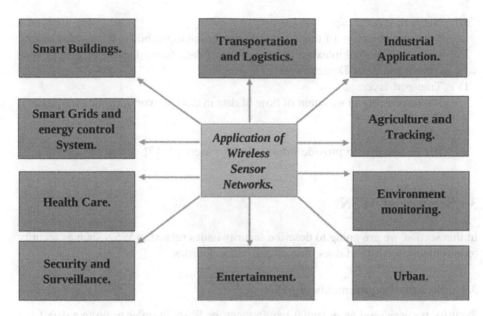

Fig. 3. Domains for WSN application [32].

Therefore, the most important fields [2, 9, 31] are:

I. Forecast temperatures and environment issues.
II. Military field, WSN applications are used for monitoring and surveillance forces of opposing and identify targets.
III. In health, it works in several fields such as measure heart pulses and monitoring patient's situation.
IV. In the business field, it used in several sectors such as delivery of products.
V. Personal objectives such as Smart home, which is used for tracking consumption for electricity, water and gas.

2.4 Communication Protocols

Identifying communication protocols will be based on WSN architecture which is considered the same of wired architecture.

I. Physical layer:
   The main aim of this layer is to increase reliability. It's responsible for collecting and transmission data. Besides that, it is responsible for establishing of communication, data rate, signal Detection, signal processing, data encryption and frequency selection. Also, it ensures communication in network, which will be either point to point or point to multi-point [9, 17].
II. Data link layer:
   The main objective of this layer is to ensure interoperability. It provides several tasks such as multiplexing of data streams, data frame Detection, access flow control and error Detection [9, 17].

III. Network layer:
   The main objective of this layer is to ensure interoperability. It provides several tasks such as multiplexing of data streams, data frame Detection, access flow control and error Detection [9, 17].
IV. Transport layer:
   It's responsible to maintain of flow of data in case networks application needs it [9, 17].
 V. Application layer:
   It's responsible to provide information for users [9, 17].

3   Security of WSN

In this section, we are going to describe security issues related to WSN such as security requirements, security classes, threats and vulnerabilities.

3.1   Security Requirements of WSN

Security is considered an essential requirement for WSN in order to protect data from unauthorized access and manipulation. Attacker looks to find vulnerabilities in order to launch an attack in order to steal data. Therefore, there are essential requirements, securing WSN.

  I. Data Confidentiality.
   Confidentiality is associated with privacy and it concentrates on hidden data from access by adversaries. Idyllically, encrypt with secret key is considered one method to hide data. This method is known by cryptography [29].
 II. Data Authentication.
   Authentication concentrates on the duration of data from accurate resource to the most accurate destination. Therefore, any illegal access should be blocked from access it [29].
III. Data Integrity.
   Integrity means, data not manipulated by malicious event once sent from one node to another. This mean, data only accessed by destination node [29].
IV. Data Availability.
   Availability means the services of Network is always available. Also, it includes network sensors are always active [4].
 V. Data Update/Fresh.
   Data updates mean, there is no any old message is replayed or used by malicious nodes [4].
VI. Access Control.
   It is associated with preventing unauthorized access to Network resources [4].

3.2   Vulnerabilities of WSN

WSN is considered vulnerable for several types of attacks due to several reasons:

I. Limited capabilities to improve security mechanisms such as encryption [17].

II. Energy consumption: usually nodes are consisted of factors which are limited to power [12].

III. Memory: nodes, memory is very small. Therefore, Operation system and Protocols should be not hold high space of memory [22].

IV. Scalability: usually each network consists of massive number of nodes. Therefore, Protocols should be able to cover a high number of devices [13].

V. Nature of nodes: nodes usually are spread into nature and should be not able to locate into harmful conditions or accessed by humans. Therefore, self-organization should be considered in the deployment process [13].

VI. Sophisticated of WSN attacks: there are several types of attacks associated for different layers. Also, there are attacks associated for multi-layers such as DoS. Besides, there are several types of attackers such as internal and external. Furthermore, there are several types of attack devices such as Laptop and Moto. While these issues indicate the vulnerabilities of WSN due to traditional security mechanism is unable to protect it against sophisticated attacks [6].

VII. The function of WSN: WSN consist of massive number of nodes. Therefore, packets sent are considered high and this needs reliable communication due to there are several functions such as routing, multi-hop...etc. [19, 35].

3.3   Attacks WSN

There are several attacks for WSN such as Jamming, DoS...etc. [13]. Also, there are several classifications of these attacks such as source of attacks, devices of attacks, behavior, processing capacity, number of associated attacks etc. However, we are going to classify attacks based onto Layer in order to identify attack effects at each layer. As mentioned before, there are five layers. Some models consist of seven layers as to include Session and Presentation Layers. However, these layers are associated with application layer. Therefore, we are going to analyze attacks based onto five layer models which includes Physical, Data Link, Network, Session/Transport, Application and multi-Layers.

1. Physical layer.
a. Jamming:
   This attacks need a powerful generator for communication signal in order to infer communication signals. In this attack, adversaries are looking to disrupt the communication of network by broadcasting a high-energy signal [12]. Therefore, a malicious node transmits signals in order to block legit access for the network. This attack needs to determine the frequency of communication. Also, it's considered as a type of DOS attack. There are different types of Jamming attack such as:
   A. Periodic: An attack sends short signal periodically in order to hamper communications [18].
   B. Trivial: An attacker sends continuous noise over a period of time in order to block whole communications [18].

C. Reactive: An attacker sends signal at the moment he recognizes that another node established transmission. This will lead to collision at another part of messages [18].

b. Radio Interference:
This attack belongs to Jamming attacks and at this attack, adversaries are looking to produces transmit signal to receiving antenna by same frequency [23].

c. Node Tempering:
In this attack, there is physical access or capture for node sensor [23].

d. Active Interference:
In this attack, an attacker tries to reply old message or manipulate the order of requested messages. This attack belongs to DOS attack and looking to disrupt or block the communication of the network. This attack depends on routing protocol and duration [18].

e. Eavesdropping:
In this attack, an attacker is looking to illegally attempt to monitor or read communications between users. Most wireless communication uses Radio frequency spectrum. Due to the communication located on Wireless technology, it could intercept. In this attack, an attacker search for confidential information from node such as private key, public key, location...etc. [18].

2. Data Link Layer.
   a. Exhaustion:
   In this attack, an attacker tries to continue retransmission till node is service and operation stopped [12]. The main techniques of this attack are retransmission, interrogation (RTS/CTS) and message modification and AcK corruption. This kind of attack is aimed to exhaust resource [25].

   b. Collision:
   In this attack, the node that comprises by hacker does not follow Protocols for Medium access control (MAC) and this will lead to generate collision with the transmission of the neighbour node by sending small noisy packet. The collision happened due to two nodes are looking to transmit through using the same frequency simultaneously. This attack made corruption, packet and disrupt the operation of networks [12]. There are several techniques associated with this attack such as change Ack message and alter field of packets [23, 25].

   c. Unfairness:
   This an attack is considered a part of DoS attack which looks to continuous exhaustion of network resources. There are several techniques associated with this attack such as applications that made collision, continuous request for access communication channel by hacker [23].

   d. Interrogation:
   In this attack, an attacker uses Request to send and Clear to send (RTS/CTS) method that provide by MAC protocols to solve problem of hidden nodes. An attacker keeps sending Request to send to target node in order to gain Clear to send [12]. This kind of attack is classified as a resource exhaustion which continuous retransmission till node became down. Also, it's considered a modification or corruption of ACK messages [23].

e. Sybil:
   In this attack, comprised node by the attacker is aimed to attract of whole network traffic after making a modification in routing. Also, one node uses multiple identities in order to get identity of the neighbour. This attack is aimed to prominent link layer and identify best link for transmission [12, 25].

3. Network Layer.
   a. Sinkhole:
   In this an attack, an attacker tries to attract traffic from regions through announce itself as an optimal path through advertise several promotions such as power, bandwidth and quality. In other words, malicious node is going to lure whole traffic or compromised node that looks more attractive for traffic [33]. Therefore, other nodes are going to consider this path as a good one. The result of this, traffic is going to be under malicious node [16]. The main effects of this an attack are Drops packets, modify information, use old data and resource consumption [26].

   b. Hello Flood:
   In this attack, Hello packets are considered the main weapon for an attacker which use it to announce itself as neighbours in order to receive connection from several nodes [5]. This attack depends sharply on neighbour information in order to establish a routing path [14]. The effects of this attack are associated for direct of packets. Therefore, it could be dropping or alter packets. Also, this attack consumes network resource due to congestion issue [26].

   c. Node Capture:
   The main aim of this an attack is to gain full control of the node. This will lead to get access to information. The effect of this attack is considered so wide due to it has control over a node in the network. Therefore, it could made alter or drop packets [5].

   d. Black Hole:
   In this attack, a malicious node promotes itself to lure whole traffic. Therefore, rest of nodes start routing traffic through a malicious node [26] while malicious node refuses route packets and dropped them. This attack, reduce throughputs dramatically of nodes near to malicious node. Besides that, it will effect traffic [16].

   e. Wormhole:
   The main aim of this an attack is to alter information, disrupt the topology and traffic flow of network [26, 28]. This an attack is created through tunnelling an old message received from a location of network and replicated it into another part of network thorough low-latency links [22]. The main effects of this an attack are traffic flow, modification of information and disrupt in network topology [26]. There are several effects of this an attack such as traffic flow, packets dropping and packet alteration, short path [30].

   f. Spoofed, Altered, or Replayed Routing Information.
   In this attack, an attacker concentrates on routing information by replying or modify routing information till network get into confusion. The main result of this type of attack is loss packet. Usually, an attacker in this attack needs to recognize traffic patterns, sources and destinations [26]. There are several

techniques associated with this an attack such as ACKs replication [22]. This an attack leads to prevent legal user to transmit information [24].

g. Misdirection:

This an attack is considered one of the most active attack. Indeed, a malicious node shows with routing path possibilities to send packets to wrong destination while this destination is considered unreachable. Also, in this attack, an attacker forward packets to wrong destination which may considered a victim node. This will lead to flood victim node and getting it into sleeping. This an attack makes consumption of resources and flood traffic [17].

h. Homing:

This an attack is designed based on traffic analysis to identify a node with especial responsibilities such as cryptographic. After identify it, an attacker can launch another attack such as DoS or jamming or order to destroy or prevent work of this node [17].

4. Session/Transport Layer.

a. Flooding:

In this an attack, an attack keeps making requests of connection till nodes are exhausted. The main effects of this an attack are exhausted node and no connection [31].

b. De-synchronization:

In this an attack, an attack designed massage for one or several end nodes, which are requesting of missed frames. The main effect of this an attack is consume energy of node [17].

c. Session Hijacking:

This attack toward TCP and UDP Protocols. In this an attack, an attacker spoof IP address of victim node. Then identify the right consequence number that is expected to get by node in order to launch an attack such as DoS [20].

d. SYN Flooding attack:

This attack belongs to DoS attack, where an attacker is looking to create several half connections with Victim through TCP without complete handshakes to entirely open connection. In TCP connection between two nodes, three-way handshake is used to establish TCP connection. While an attack is running, a malicious node sends massive number of SYN packets to target node and spoofs the return address from SYN packets. Target node sends SYN-ACK packets after got SYN packets from an attacker and it wait ACK packets. Without getting ACK packets, data of half-open connection will stay in target node. In case, these connections stored in a fixed size of table while waiting ACK, all pending connections can consume buffer. Therefore, there are no any possibilities for any new connection [20].

5. Application Layer.

a. Overwhelm:

In this attack, an attacker is looking to overwhelm network sensors through generating large volumes of traffic toward base station. This will drain the energy of system and network bandwidth [11, 12].

b. Path-based DOS:

In this attack, adversaries are exhausted sensors by flood end-to-end communication path by either injected message or replay in order to consume energy [1].

c. Malicious code:

Malicious code presents Viruses, spyware, Trojan and worms. Application Layer is dealing with the operating system (OS) and User application. Besides that, it uses several Protocols such as IP, TCP, UDP, SMTP...etc. Therefore, these malicious code can duplicate their-self and spread into the system. This will make the system slow down and even could damage the operating system [20].

6. Multi-layers.

a. Denial-of-Services (DoS):

The main aim of this an attack is consuming power, memory, bandwidth and disrupt the network [10]. Also, it reduces network availability [25]. This an attack can be launched at several layers such as jamming and Tempering at the physical layer while at link layer, malicious, collision and Exhaustion control communication channel through capture traffic. At the network layer, misdirection or routing information could be disrupted, dropped or modified. At Transport layer, Desynchronization and Flooding attack and Application layers such as Flooding, SYN, Malicious node and Session Hijacking [1].

b. Man-in-the-Middle:

In this an attack, an attack located between sender and receiver in order to eavesdropping information between both of them. In some cases, an attacker impersonate as sender to get information from receiver or sometime impersonate as receiver to get information from receiver [33].

c. Impersonation:

In this attack, an attacker use identity for node such as IP or MAC address. Usually, this attack is associated with another attack [20].

4   WSN Metrics

In this section, we are going to identify measure metrics for network behavior. Also, we are going to propose a model that reflect network behavior. Table 1, summarized attacks and effects into each layer.

Table 1, shows several impacts for attacks. In Table 2, we identified metrics to measure network behaviour based on attacks effects.

1. Services Metrics.

Defined Metrics for measure services has two dimensions, layer-service orientation [8] Table 3 and Quality of Services (QoS) Table 4 [27]. Besides that, there are two perspectives, firstly user and application perspective. Secondly, network perspective. In the first one, concentration on quality of application while the concentrating on the second one is quality of application with efficient utilization of network [8].

Table 1. Effect of attacks into each layer [1, 5, 10–12, 15–17, 19, 21, 22, 24, 25, 27, 29, 30, 32].

Attacks.	Layers.	Functional Effects.	QoS Effects.
1. Radio Interference/ Jamming. 2. Tampering. 3. Path-based DoS (PDoS). 4. Node outage.	Physical	1. Radio interference, resource exhaustion, block legit access. 2. Damage node, stop or change service. 3. Power consumption, network disruption, network services. 4. Stop services	1. Availability. 2. Bandwidth. 3. Throughput. 4. Delay.
1. Exhaustion. 2. Unfairness. 3. Interrogation. 4. Sybil.	Data Link.	1. Stop services and operations. 2. Packet corruption and disrupt of network operation. 3. Exhaustion of network services. 4. Attract network traffic.	1. Availability. 2. Throughput. 3. Delay. 4. Loss. 5. Error rate. 6. Traffic.
1. Sinkhole 2. Hello Flood. 3. Node Capture. 4. Black Hole. 5. Wormhole 6. Spoofed, Altered or Replayed Routing Information 7. Misdirection. 8. Homing.	Network.	1. Attract traffic, drop packets, resource Consumption 2. Drop or alter packets, consume network resources. 3. Node control. 4. Lure traffic, routing, drop packet, reduce throughput. 5. Traffic flow, drop and alter packets. 6. Loss packets, prevent legal users to transmit. 7. Resource consumption and flood traffic. 8. Lunch another attacks such as DoS or Jamming in order to destroy or prevent functionality of node.	1. Availability. 2. Throughput. 3. Delay. 4. Loss. 5. Error rate. 6. Services. 7. Connection.
1. Flooding 2. De-synchronization 3. Session hijacking 4. SYN Flooding attack.	Transport	1. Exhausted node. 2. Exhausted node. 3. Spoofs IP addresses. 4. Buffer and connection consumption.	1. Availability. 2. Throughput. 3. Connection. 4. Loss.
1. Overwhelm 2. Path-based DoS. 3. Malicious code.	Application	1. Consume resources and bandwidth of network. 2. Flood communication. 3. Slow down system, could damage Operation System.	1. Availability. 2. Throughput. 3. Delay. 4. Loss. 5. Error rate. 6. Bandwidth.
1. Denial-of-Services (DoS). 2. Man-in-the-Middle.	Multi-Layers	1. Consume Power, memory and bandwidth, disrupt network.	1. Availability. 2. Throughput. 3. Delay. 4. Loss. 5. Error rate.

Table 2. Measurement metrics for network behavior.

Measurements	Layers
Services	Physical
Bandwidth	
Connection	
Resources and Power	
Services	Data Link
Alter messages	
Retransmission	
Packet corruption	
Resources	
Network Traffic	
Traffic	Network
Drop packets	
Resources	
Alter messages	
Control of Node	
Throughput	
Traffic Flow	
Transmit information	
Prevent Nodes to work	
Resources (Buffer and Connections)	Transport
Spoof IP address	
Resources	Application
Bandwidth	
Flood communication	
Operating System	

Table 3. Layer service metrics [8].   **Table 4.** Quality of service metrics [27].

Layer Service Metrics.	
Layer	Metrics Parameters
Application	System Lifetime
	Response time
Transport	Reliability
	Bandwidth
	Latency
Network	Capacity
	Connectivity
	Throughput
	Energy

Quality of Services (QoS).
Availability
Error rate
Loss
Bandwidth
Throughput
Delay
Jitter

2. Connection and Fairness metric.

This metric adapted from experiments using Snort and Wireshark Software Table 5.

Table 5. Communication and fairness metric.

Connection and Fairness Metrics.
Source Address
Destination Address
Source Port
Destination Port
Type of Protocol
Packets
Time

3. Resource and Traffic metrics.

   These two metric are adapted from experiment using Wireshark and Snort Software. Tables 6 and 7 show both of them respectively.

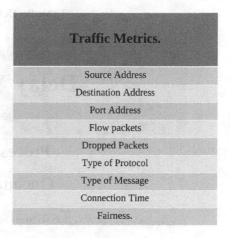

Table 6. Resource metric.

Resources Metrics.
Source Address
Destination Address
Source Port
Destination Port
Memory
CPU
Capacity
Utilization
Fairness

Table 7. Traffic metric.

Traffic Metrics.
Source Address
Destination Address
Port Address
Flow packets
Dropped Packets
Type of Protocol
Type of Message
Connection Time
Fairness.

4. Fragmentation metric.

   This metric adapted based on experiments on Wireshark Software Table 8.

Table 8. Fragmentation metric.

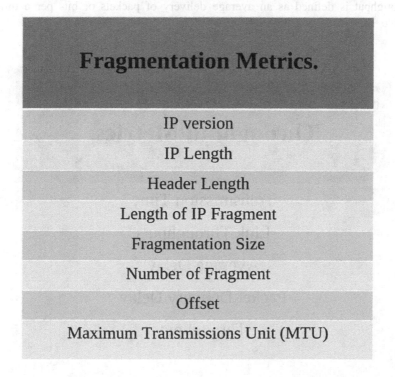

Fragmentation Metrics.
IP version
IP Length
Header Length
Length of IP Fragment
Fragmentation Size
Number of Fragment
Offset
Maximum Transmissions Unit (MTU)

5. Delay metric.
   Delay defined as an average time between sending and receiving packets [7] (Table 9).

Table 9. Delay metrics [3].

Delay Metrics.
Process Delay
Queuing Delay
Transmission Delay
Propagation Delay

6. Throughput Metric.
   Throughput is defined as an average delivery of packets or bits per a time [3] (Table 10).

Table 10. Throughput metric [3].

Throughput Metrics.
Transmission Time
Link Throughputs
Average Delay
Packet Delivery Delay
Throughput

7. Availability Metric.
   Availability definition is an ability to do a task by functional unit and it is associated with connectivity and functionality [3]. Therefore, measure availability will be based on whole previous metrics together concerning service continuity.

5 Results Models

This section included generated models that related to WSN.

1. Figure 6, shows proposed Security model of WSN into Framework of Security Architecture of Smart City based onto Intrusion Detection (IDs).
2. Figure 4a and b, shows proposed model about security of WSN.
3. Figure 5, shows a proposed model about WSN Metrics.

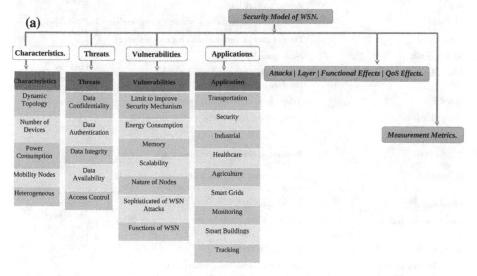

Fig. 4. (a) Proposed security model for WSN. (b) Proposed security model for WSN. (Belong for Fig. 4a.)

(b)

Security Model of WSN.

Attacks | Layer | Functional Effects | QoS.

Measurement Metrics.

Attacks.	Layers.	Functional Effects.	QoS Effects.
Radio Interference/ Jamming.	Physical	Radio interference, resource exhaustion, block legit access	1. Availability. 2. Bandwidth. 3. Throughput. 4. Delay.
Tampering.		Damage node, stop or change service.	
Path-based DoS (PDoS).		Power consumption, network disruption, network services.	
Node outage.		Stop services	
Exhaustion.	Data Link.	Stop services and operations.	1. Availability. 2. Throughput. 3. Delay. 4. Loss. 5. Error rate. 6. Traffic.
Unfairness.		Packet corruption and disrupt of network operation.	
Interrogation.		Exhaustion of network services.	
Sybil.		Attract network traffic.	
Sinkhole	Network.	Attract traffic, drop packets, resource Consumption	1. Availability. 2. Throughput. 3. Delay. 4. Loss. 5. Error rate. 6. Services. 7. Connection.
Hello Flood.		Drop or alter packets, consume network resources.	
Node Capture.		Node control.	
Black Hole.		Lure traffic, routing, drop packet, reduce throughput.	
Wormhole		Traffic flow, drop and alter packets.	
Spoofed, Altered or Replayed Routing Information		Loss packets, prevent legal users to transmit.	
Misdirection.		Resource consumption and flood traffic.	
Homing.		Lunch another attacks such as DoS or Jamming in order to destroy or prevent functionality of node.	
Flooding	Transport	Exhausted node.	1. Availability. 2. Throughput. 3. Connection. 4. Loss.
De-synchronization		Exhausted node.	
Session hijacking		Spoofs IP addresses.	
SYN Flooding attack.		Buffer and connection consumption.	
Overwhelm	Application	Consume resources and bandwidth of network.	1. Availability. 2. Throughput. 3. Delay. 4. Loss. 5. Error rate. 6. Bandwidth.
Path-based DoS.		Flood communication.	
Malicious code.		Slow down system, could damage Operation System.	
Denial-of-Services (DoS).	Multi-Layers	Consume Power, memory and bandwidth, disrupt network.	1. Availability. 2. Throughput. 3. Delay. 4. Loss. 5. Error rate. 6. Bandwidth.
Man-in-the-Middle.			

Fig. 4. (*continued*)

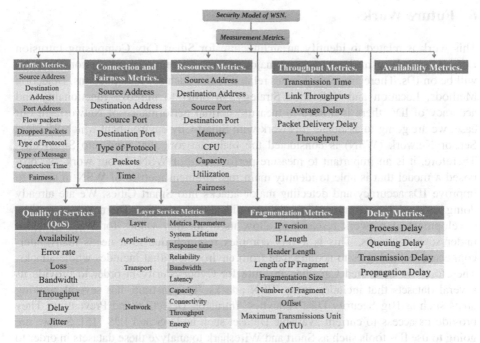

Fig. 5. Proposed model for WSN metrics (Belong for Fig. 4a.)

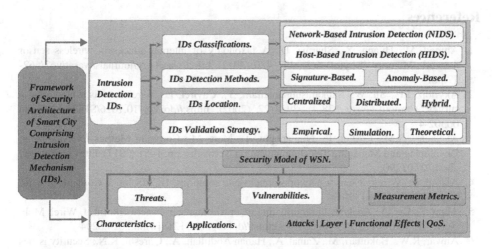

Fig. 6. Framework of security architecture of Smart City based onto (IDs).

6 Future Works

This work is related to identify an architecture for Smart City Comprising Intrusion Detection mechanism (IDs) to detect major attacks. Therefore, the main concentration will be on IDs. There are several issues related to IDs such as Classifications, Detection Methods, Location and Validation Strategy. In order to improve functionality and accuracy of IDs, there is a need to measure normal performance of Network. In our case, we are going to associate this work with Smart City environment which Wireless Sensor Network (WSN) is considered the main network that used into Smart city. Therefore, it is an important to measure performance of WSN. In our work, we proposed a model that is able to identify main measurement metrics of WSN in order to improve IDs accuracy and detecting major attacks into Smart Cities. We are already doing tests in lab environment in order to validate these metrics. The next step, will be development of network research to allow us to identify values for WSN behavior under several attacks. This is a very complex research due to it needs many interconnected technologies to acquire data from environment that includes whole attacks. Therefore, we contacted Canadian Institute for Cybersecurity in order to provide us several datasets that includes many WSN attacks. This institute has several working areas such as Big Security Data Analytics, Intrusion Detection and Prevention. They provide us access to current available Dataset such as DoS and IDS [15]. So, we are going to use IDs tools such as Snort and Wireshark to analyze these datasets in order to fulfil purpose another research.

References

1. Ahmed, M., Huang, X., Sharma, D.: A taxonomy of internal attacks in wireless sensor network. Memory (Kbytes) 6(2), 427–430 (2012). http://waset.org/journals/waset/v62/v62-77.pdf
2. Akyildiz, I.F., Su, W., Sankarasubramaniam, Y., Cayirci, E.: Wireless sensor networks: a survey. Comput. Netw. 38(4), 393–422 (2002). https://doi.org/10.1016/S1389-1286(01)00302-4
3. Al-Shehri, S.M., Loskot, P., Numanoglu, T., Mert, M.: Common Metrics for Analyzing, Developing and Managing Telecommunication Networks, pp. 1–51 (2017). https://arxiv.org/abs/1707.03290
4. Alajmi, N.: Wireless sensor networks attacks and solutions. IJCSIS Int. J. Comput. Sci. Inf. Secur. 12(7), 37–40 (2014)
5. Alam, S., De, D.: Analysis of security threats in wireless sensor network. Int. J. Wirel. Mob. Netw. (IJWMN) 6(2), 35–46 (2014). https://doi.org/10.5121/ijwmn.2014.6204
6. Anwar, R.W., Bakhtiari, M., Zainal, A., Hanan Abdullah, A., Qureshi, K.N.: Security issues and attacks in wireless sensor network. World Appl. Sci. J. 30(10), 1224–1227 (2014). https://doi.org/10.5829/idosi.wasj.2014.30.10.334
7. Balen, J., Martinovic, G., Hocenski, Z.: Network performance evaluation of latest windows operating systems. In: 2012 20th International Conference on Software, Telecommunications and Computer Networks (SoftCOM), pp. 1–6 (2012). http://ieeexplore.ieee.org/stamp/stamp.jsp?arnumber=6347604

8. Balen, J., Zagar, D., Martinovic, G.: Quality of service in wireless sensor networks: a survey and related patents. Recent Patents Comput. Sci. **4**(3), 188–202 (2011). https://doi.org/10.2174/2213275911104030188

9. Bokare, M., Ralegaonkar, A.: Wireless sensor network. Int. J. Comput. Eng. Sci. (IJCES) **2** (3), 55–61 (2012). http://vixra.org/pdf/1208.0129v1.pdf

10. Bonguet, A., Bellaiche, M.: A survey of Denial-of-Service and distributed Denial of Service attacks and defenses in cloud computing. Future Internet **9**(3) (2017). https://doi.org/10.3390/fi9030043

11. Chelli, K.: Security issues in wireless sensor networks: attacks and countermeasures. In: Proceedings of the World Congress on Engineering 2015, WCE 2015, London, U.K., vol. I, 1–3 July 2015. http://www.iaeng.org/publication/WCE2015/WCE2015_pp519-524.pdf

12. Diaz, A., Sanchez, P.: Simulation of attacks for security in wireless sensor network. Sensors **16**(11), 1932 (2016). https://doi.org/10.3390/s16111932

13. Garcia-Font, V., Garrigues, C., Rifà-Pous, H.: Attack classification schema for smart city WSNs. Sensors (Switzerland) **17**(4) (2017). https://doi.org/10.3390/s17040771

14. Gill, R.K., Sachdeva, M.: Detection of hello flood attack on LEACH in wireless sensor networks. Adv. Intell. Syst. Comput. **638**, 377–387 (2018). https://doi.org/10.1007/978-981-10-6005-2_40

15. Habibi, A.: Candadian Institute for Cybersecurity (CIC), University of New Brunswick (UNB), Frederication, Canada (2018). http://www.unb.ca/cic/about/index.html

16. Kaur, D., Singh, P.: Various OSI layer attacks and countermeasure to enhance the performance of WSNs during wormhole attack. ACEEE Int. J. Netw. Secur. **5**(1) (2014). http://searchdl.org/public/journals/2014/IJNS/5/1/11.pdf

17. Kavitha, T., Sridharan, D.: Security vulnerabilities in wireless sensor networks: a survey. J. Inf. Assur. Secur. **5**(2010), 31–44 (2010)

18. Khatri, S.: A taxonomy of physical layer attacks in MANET **117**(22), 6–11 (2015)

19. Liu, Y.M., Wu, S.C., Nian, X.H.: The architecture and characteristics of wireless sensor network. In: 2009 International Conference on Computer Technology and Development, ICCTD 2009, vol. 1, no. 561, pp. 561–565 (2009). https://doi.org/10.1109/ICCTD.2009.44

20. Lupu, T., Parvan, V.: Main types of attacks in wireless sensor networks. In: Proceedings of WSEAS International Conference. Recent Advances in Computer Engineering, pp. 180–185 (2009)

21. Miorandi, D., Sicari, S., De Pellegrini, F., Chlamtac, I.: Internet of things: vision, applications and research challenges. Ad Hoc Netw. **10**(7), 1497–1516 (2012). https://doi.org/10.1016/j.adhoc.2012.02.016

22. Mohammadi, S.: A comparison of link layer attacks on wireless sensor networks. J. Inf. Secur. **02**(02), 69–84 (2011). https://doi.org/10.4236/jis.2011.22007

23. Mohammadi, S., Jadidoleslamy, H.: A comparison of physical attacks on wireless sensor networks. Int. J. Peer to Peer Netw. **2**(2), 24–42 (2011). https://doi.org/10.5121/ijp2p.2011.2203

24. Yılmaz, M.H., Arslan, H.: A survey: spoofing attacks in physical layer security. In: Proceedings - Conference on Local Computer Networks, LCN, December 2015, pp. 812–817 (2015). https://doi.org/10.1109/LCNW.2015.7365932

25. Nandal, V.: Comparison of attacks on wireless sensor networks **3**(7), 208–213 (2014)

26. Patel, M., Aggarwal, A., Chaubey, N.: Wormhole attacks and countermeasures in wireless sensor networks : a survey. Int. J. Eng. Technol. **9**(2), 1049–1060 (2017). https://doi.org/10.21817/ijet/2017/v9i2/170902126

27. Pereira, V.N.S.S.: Performance measurement in wireless sensor networks, p. 213 (2016)

28. Pongle, P., Chavan, G.: Real time intrusion and wormhole attack detection in internet of things. Int. J. Comput. Appl. **121**(9), 975–8887 (2015). https://doi.org/10.5120/21565-4589

29. Kirar, V.P.S.: A survey of attacks and security requirements in wireless sensor networks. World Acad. Sci. Eng. Technol. Int. J. Electron. Commun. Eng. **8**(12) (2014). https://waset.org/publications/10000089/a-survey-of-attacks-and-security-requirements-in-wireless-sensor-networks

30. Raote, N.S.: Defending wormhole attack in wireless ad-hoc network. Int. J. Comput. Sci. Eng. Surv. **2**(3), 143–148 (2011). https://doi.org/10.5121/ijcses.2011.2311

31. Rghioui, A., Bouhorma, M.: 6lo technology for smart cities development : security case study **92**(15), 54–59 (2014). https://doi.org/10.5120/16089-5402

32. Shahzad, F., Pasha, M., Ahmad, A.: A Survey of Active Attacks on Wireless Sensor Networks and their Countermeasures, vol. 14, no. 12, pp. 54–65 (2017). http://arxiv.org/abs/1702.07136

33. Subramanian, V.: Proximity-based attacks in wireless sensor networks. Int. J. Sci. Eng. Res. **3**(May 2013), 2–5 (2013)

34. Tellez, M., El-Tawab, S., Heydari, M.H.: IoT security attacks using reverse engineering methods on WSN applications. In: 2016 IEEE 3rd World Forum on Internet of Things (WF-IoT), pp. 182–187 (2016). https://doi.org/10.1109/WF-IoT.2016.7845429

35. Zarpelão, B.B., Miani, R.S., Kawakani, C.T., de Alvarenga, S.C.: A survey of intrusion detection in internet of things. J. Netw. Comput. Appl. (2017). https://doi.org/10.1016/j.jnca.2017.02.009

Attacking Strategy of Multiple Unmanned Surface Vehicles Based on DAMGWO Algorithm

Juan Pu[1], Xing Wu[1,2(✉)], Yike Guo[1], Shaorong Xie[3],
Huayan Pu[3], and Yan Peng[3]

[1] School of Computer Engineering and Science,
Shanghai University, Shanghai, China
xingwu@shu.edu.cn
[2] Shanghai Institute for Advanced Communication and Data Science,
Shanghai, China
[3] School of Mechatronic Engineering and Automation,
Shanghai University, Shanghai, China

Abstract. Unmanned combat system has received more and more attention with the development of modern weapons and equipment in recent years, which results in the application of unmanned surface vehicles (USVs) in the military. The USVs are designed to attack the protected hostile target with a minimum loss that use the lower attack capacity and a small number of attack USVs to reach the target. The USVs attack problem could be viewed as a multi-constrained task assignment problem. In this study, a novel algorithm is proposed, which is called DAMGWO, a grey wolf optimization (GWO) algorithm based on distributed auction mechanism (DAM). This algorithm combines DAM and GWO algorithm to constrain the wolf initialization, increasing the ability to break away from the local optimum. Furthermore a corresponding fitness function to evaluate the quality of this algorithm is proposed. The experimental results show that the proposed algorithm not only fully meets the requirements of the attacking strategy of multiple USVs that to attack the protected hostile target with a minimum loss, but also has better convergence than the traditional GWO algorithm and PSO algorithm.

Keywords: Unmanned Surface Vehicles (USVs) · Attack strategy
Grey wolf optimization · Distributed auction mechanism · Task assignment

1 Introduction

With the development of science and technology, especially the innovation of information technology, a lot of new material technology, new military theories, new ideas, new methods and new weapons have emerged that subvert the tradition [1]. Unmanned combat system is a new type of combat system which can replace people to accomplish a specific task, and it is the product of the comprehensive integrated innovation and development of a series of advanced military "technical groups". Nowadays, it has been widely used in ocean, land and air, such as unmanned aerial vehicle platform,

© Springer Nature Switzerland AG 2018
S. Wrycza and J. Maślankowski (Eds.): SIGSAND/PLAIS 2018, LNBIP 333, pp. 135–145, 2018.
https://doi.org/10.1007/978-3-030-00060-8_10

ground operations, underwater unmanned vehicle platform operations of unmanned aircraft platform and surface warfare unmanned craft platform (Unmanned Surface Vehicles, USV) [2, 3].

Compared with the surface ship, USV has the characteristics of flexibility, good concealment, wide activity area and low using cost [4]. The biggest advantage of the USV is to avoid the risk of casualties of the naval personnel and to accomplish tasks that cannot be completed under human conditions. It can be seen that the development of USV is very urgent in civilian and military fields.

The biological cluster and competitive advantages in nature are shown in Fig. 1. In terms of the advantages of individuals and cluster, we can use the way of wolf attack to solve other relevant problems. The wolf attack is a behavior which is produced in the process of hunting in nature, such as wolves searching behavior, wolves besiege behavior and so on [5].

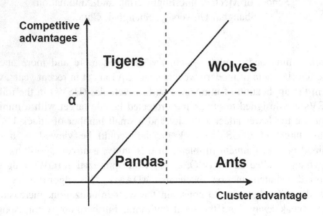

Fig. 1. Cluster and competitive advantages of nature groups

Inspired from the view of bionic, we apply wolf attack into USVs combat. Wolf attack uses the power of vehicles to attack one marine target based on the Wolf-pack. In the process of the attack, we can use the threat assessment methods to assess the threat index of both sides [6].

In this paper, we investigate some representative articles for the development of USV in Sect. 2 and introduce the main theory of specific attack strategy in Sect. 3. Then the experiment setting and results are put in Sect. 4. Finally, we conclude this paper in Sect. 5.

2 Related Work

Task planning of the joint attack among Multiple USVs is the key to ensure successfully complete tasks. Multiple USVs need to take some constraints into account, such as the environmental factors, the complexity of different tasks and the capacity to perform various complex attack task in the execution of joint attacks. When dealing with the above problems, the traditional attack task scheduling methods do not take into account the external and internal constraints of the attack task, but simply decomposes the complex attack task into the superposition of the individual. In the process of task execution, we can not only achieve the purpose of improving the task execution efficiency, but also avoid falling into chaos due to the lack of corresponding collaborative planning mechanism. Task planning is to solve the optimization problem of using the minimum loss to get the highest benefit when Multiple USVs cooperate to execute tasks [7].

Some scholars have used genetic algorithms [8], ant colony algorithms [9], contract networks [10] and other methods to solve the similar problems. However, the previous studies did not take into account the multiple constraints of the Multiple USVs task planning problems [11]. On this basis, some scholars have used Particle Swarm Optimization (PSO) [12] and its improvement to solve multiple USVs task planning problems [13]. It regards the cluster attack task planning problem as multi-constrained task assignment process and builds task planning model. A PSO optimization algorithm based on distributed auction mechanism is proposed to improve particle initialization and optimization process, so that it can meet task constrain condition and maintain diversity.

This paper uses a GWO [14] algorithm based on DAM [15] to study attack strategy of Multiple USVs. The results of the study are finally compared with other swarm intelligence algorithms.

3   Multiple USVs Attack Strategy Model and Method

In the biological world, wolves cooperative hunting embodies the advantages of cluster. As early as the Second World War, wolves tactics were used by the German Navy. Its essence is to destroy a large fleet by concentrating the forces of many small ships. Here we use the idea of wolves cooperative hunting in USVs operations. The ultimate aim is to attack the protected hostile target and accomplish the tasks with the minimum loss. And the protect-attack diagram of USVs as shown in Fig. 2:

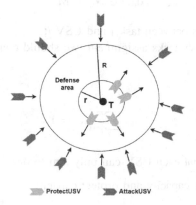

Fig. 2. The protect-attack diagram of USVs

The Attack USVs advance towards the target from different directions with a same velocity, the protection USVs are in the circumference area where the distance from radius of target is r (Target location has been set in advance as an environmental assumption). When attack USVs enter into the circumference area which the distance from radius of target is R, two kinds of USVs can detect each other. System console and information processing equipment are used to collect the coordinates, velocity, attack capacity and so on.

3.1 Multiple USVs Attack Strategy Model and Planning Function

Assuming a multiple USVs group is performing tasks in a two-dimensional space, the sea environment, threatening region and the target location have been obtained in advance, thus multiple USVs task assignment problems can be represented as follows: E represents sea environment and threaten region, N represents attack USV's information, M represents the information of tasks, T represents target's information, A represents USV's attack capacity, P represents attack capacity required by task, that is, the protection capacity of tasks.

According to the above description, We define several decision variables as follow:

$$x_{ij} = \begin{cases} 1 \\ 0 \end{cases}, i \in N, j \in M \tag{1}$$

Where $x_{ij} = 1$ represents that task j is acted by USV i, in other cases, $x_{ij} = 0$;

$$a_i, i \in N \tag{2}$$

Where a_i is attack capacity of USV i;

$$p_j, j \in M \tag{3}$$

Where p_j is defensive capacity of task j;

$$d_{ij}, i \in N, j \in M \tag{4}$$

Where d_{ij} is the distances between task j and USV i;

In the process of USVs tasks assignment, we should consider the following constraint conditions:

(1) Constraint on USV

$$\sum_{j=1}^{m} x_{ij} = 1 \tag{5}$$

Formula (5) represents that each USV can only aim to one task;

(2) Constraint on attack capacity and protect capacity

$$\sum_{i=1}^{n} a_i x_{ij} \geq p_j \tag{6}$$

Formula (6) represents that the sum of attack capacity of executing task j must satisfy the required attack capacity of accomplishing task j.

According to the definition of decision variables and confinement problems above, the multiple USVs task planning in attack strategy can be described as follow. According to the present environment, tasks and the distributing situation, multiple USVs task assignment problem is the optimal problem to make the minimum loss for executing all tasks. Fitness function for solving this problem can be expressed as:

$$f = \min \left[\sum_{i=0}^{n} \sum_{j=0}^{m} x_{ij} d_{ij} + \sum_{i=0}^{n} \sum_{j=0}^{m} x_{ij} + \left(\sum_{i=0}^{n} \sum_{j=0}^{m} a_i x_{ij} - \sum_{j=0}^{m} p_j \right) \right] \tag{7}$$

Formula (7) represents the highest benefit of USVs to reach the target with the minimum loss.

3.2 Problem Description of Multiple USVs Attack Strategy

Grey wolf optimization (GWO) was first introduced by Mirjalili et al. [14]. It is a swarm intelligence algorithm to simulate the predatory behavior of wolves. According to the social hierarchy of grey wolves, the algorithm simulates the predatory behavior and hierarchy of grey wolves in nature. In order to mathematically model the social hierarchy of grey wolves when designing GWO, the fittest solution is considered as alpha (α). Consequently, the second and third best solutions are named as beta (β) and delta (δ), respectively. The rest of the candidate solutions are assumed to be omega (ω). In the GWO algorithm, the hunting (optimization) is guided by α, β, and δ and the ω wolves follow these three wolves.

Assuming that there are N attack USVs, and M defense tasks around the target. Then the task assignment matrix is $X = [x_{ij}]_{n \times m}$, x_{ij} represents the USV which is assigned by the master to execute the task j.

$$X = (X_1, X_2, \cdots, X_n) = \begin{bmatrix} x_{11} & \cdots & x_{1m} \\ \vdots & \ddots & \vdots \\ x_{n1} & \cdots & x_{nm} \end{bmatrix} \tag{8}$$

Then $\sum_{i=1}^{n} x_{ij}, j = 1, 2, \cdots m$ represents the amount of USVs which are assigned by master to execute task j. And the $\sum_{j=1}^{m} \sum_{i=1}^{n} x_{ij}, i = 1, 2, \cdots n, j = 1, 2, \cdots m$, represents the amount of USVs which are assigned by master to execute all tasks. Attack capacity matrix is $A = [a_1 a_2 \cdots a_n]^T$, a_i, $i = 1, 2, \cdots n$, represents attack capacity of USV i. Protection capacity matrix is $P = [p_1 p_2 \cdots p_m]^T$, p_j, $j = 1, 2, \cdots m$, represents protection capacity of task j, and it is the attack capacity of accomplishing task j.

The following two constraints need to be considered when establishing mathematical model based on practical problem:

(1) Attack capacity constraint condition: the attack capacity of each task must be equal or greater than the needed attack capacity $P \leq X \times A$.

$$
\begin{bmatrix} p_1 \\ p_2 \\ \vdots \\ p_m \end{bmatrix} \leq \begin{bmatrix} x_{11} & \cdots & x_{1m} \\ \vdots & \ddots & \vdots \\ x_{n1} & \cdots & x_{nm} \end{bmatrix} \times \begin{bmatrix} a_1 \\ a_2 \\ \vdots \\ a_n \end{bmatrix} \tag{9}
$$

In formula (9), $p_j = \sum_{i=0}^{n} a_i x_{ij}$.

(2) The number of the attack USVs constraint condition: the number of the attack USVs assigned by master for tasks must equal or less than the total number of attack USVs:

$$
\sum_{j=1}^{m} \sum_{i=1}^{n} x_{ij} \leq n, i = 1, 2, \cdots n, j = 1, 2, \cdots m \tag{10}
$$

From the above, in the GWO algorithm, the pursuit behavior is carried out by α, β and δ. The ω follows them to track and suppress the prey, and then completes the predation task. When using GWO algorithm to solve the optimization problem, in the d-dimension search space, we assume that the number of grey wolf individuals is N in the population of wolves. The position of wolf i in d-dimensional space can be expressed as $x_i = x_{i1}, x_{i2}, x_{i3}, x_{i4}, \ldots, x_{id}$, the current optimal individual in the population is recorded as α, the current suboptimal individual is recorded as β, and the third one is δ, then the others are ω, the prey's position corresponds to the global optimal solution of the optimization problem. And then the optimization process of the GWO algorithm is: A group of grey wolf individuals are randomly generated in the search space. For this group, fitness evaluation is performed to obtain the top three grey wolf individuals α, β, δ, which serves as a benchmark for finding the position of the prey (global optimal solution). The position of the next generation of grey wolf individuals are calculated based on α, β and δ.

In the process of preying, grey wolves encircle prey during the hunt. Corresponding to the optimization process of the GWO algorithm, the distance between the prey and the grey wolves can be expressed as:

$$
D = |C \cdot X_P(t) - X(t)| \tag{11}
$$

$$
X(t+1) = X_P(t) - A \cdot D \tag{12}
$$

Where t is the current iteration, $X_P(t)$ is the position vector of the prey, and $X(t)$ is the position vector of a grey wolf. A and C are coefficient vectors, and they are calculated as follows:

$$
A = 2a \cdot r_1 - a \tag{13}
$$

$$
C = 2 \cdot r_2 \tag{14}
$$

Where a decreases linearly from 2 to 0 as the number of iteration increases. and r1, r2 are random vectors in [0, 1].

After encircling the prey, the ω wolf chased the prey with the lead of the α, β and δ. During the hunt, the position of the wolf group is changed with the prey, and it could be redefined according to the position of the α wolf.

$$\begin{cases} D_\alpha = |C_1 X_\alpha(t) - X(t)| \\ D_\beta = |C_2 X_\beta(t) - X(t)| \\ D_\delta = |C_3 X_\delta(t) - X(t)| \end{cases} \tag{15}$$

$$\begin{cases} X_1 = X_\alpha(t) - A_1 D_\alpha \\ X_2 = X_\beta(t) - A_2 D_\beta \\ X_3 = X_\delta(t) - A_3 D_\delta \end{cases} \tag{16}$$

$$X(t + 1) = \frac{X_1 + X_2 + X_3}{3} \tag{17}$$

Where D_α, D_β and D_δ are the distance between α, β and δ wolves and ω wolves (other individuals) respectively.

Attacking is the last stage of wolves hunting. The wolf group attacks the prey and captures the prey, which means that the optimal solution is obtained. The realization of the process is mainly achieved by decrement of a value. GWO has only one main parameter (a) to be adjusted. The adaptive values of parameters a and A allow GWO to smoothly transition between exploration and exploitation. Therefore, exploration and exploitation are guaranteed by the adaptive values of a and A. With decreasing A, half of the iterations are devoted to exploration ($|A| \geq 1$) and the other half are dedicated to exploitation ($|A| < 1|$).

3.3 Improvement of GWO by Adding DAM in Algorithm Initialization (DAMGWO)

In the process of grey wolf initialization, it should satisfy not only the constraint conditions of attack capacity and the USV number, but also the constraint conditions of USV's attacking radium. Without considering the conditions above, the task assignment matrix generated by random initialization is difficult to satisfy the requirement in practical problems.

Traditional GWO has the following problems in solving problem of multiple USVs attack strategy [13, 16]:

(1) Due to the constraints of the USV number and attack capacity, randomly generated initial wolves are difficult to satisfy the requirements in practical problems.
(2) The large number of USVs participating in task assignment makes a large dimension, just using traditional algorithm makes the process longer, and the algorithm in the optimization process is easy to fall into local optimum.

According to above two problems, the initial state of the wolf group is constrained by adding DAM algorithm to the initial wolves. Here is the introduction of the specific optimization steps as follow: initialization, task announcement, bidding, awarding, executing and finish.

- Step 1 Initialization: Set the GWO algorithm parameters.
- Step 2 Task Announcement: The master (multiple USVs control terminal) will detect and get the tasks information, and then send it to each USV individual by the net.
- Step 3 Bidding: The idle USVs first calculate whether they had feasible path and executive capacity to perform this task after receiving the bidding information, then a bidding value can be calculated and returned according to themselves properties, such as position, constraints and attack capacity.
- Step 4 Awarding: The tenderer deals with the bidding after receiving bidding values of all idle USVs or reaching the bidding deadline, and the best bidding will be selected by the predetermined standard. Then it will send out the bidding message to the best bidding USV that put forward the best bidding and the failed message to the others.
- Step 5 Executing: According to the result of bidding, the master get a globally optimal solution as the initial value to initial wolves.
- Step 6 Individual Evaluation: The fitness values of all wolves are evaluated by the GWO algorithm, and the current three optimal solutions are obtained to update the position of the current wolf.
- Step 7 Finish: Judging whether the optimal solution has been obtained, or it has reached the max iteration number. If not, updating wolves and return to step 3. As shown in Fig. 3.

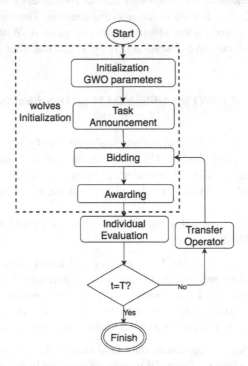

Fig. 3. Flow chart of DAMGWO algorithm

4 Experiments and Results

In order to verify the superiority of the algorithm, a simulation analysis is carried out by an example. In the simulation experiment, we set the parameters as shown in Table 1. The target position is (0, 0), and we use 10 USVs to attack 8 different protection tasks from different randomly positions. The protection security range radius is 20 m, and the alarm range radius is 50 m. And then the number of iterations is 100.

Table 1. Parameters for DAMGWO algorithm

Parameter	Value
Population size	50
Maximum number of iterations	100
Target position	(0, 0)
Attack USVs	10
Protection USVs	8
Defense radius	r = 20
Defense alert range	R = 50

In order to show the superiority of DAMGWO algorithm in solving this problem under constraints, we use the traditional GWO algorithm and the PSO algorithm for comparative experiments. The setting of the experimental parameters are shown in Table 2.

Table 2. Parameters for the three algorithms

Algorithm	Population size	Maximum number of iterations
GWO	50	100
DAMGWO	50	100
PSO	50	100

The simulation results are shown in Fig. 4. It can be obviously found that the results of GWO algorithm is not the global optimal solution. According to the practical situation, the DAMGWO algorithm formed by the improved GWO algorithm obtains a better global optimal solution. The PSO algorithm falls into a local optimum. Because of the constraints, there is not a good diversity of the PSO algorithm [13]. The particles become scarce during updating, which leads to the algorithm falling into a local optimum, and then it is unable to obtain a global optimal solution.

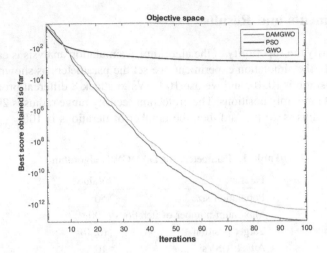

Fig. 4. The curve of the optimal solution of the problem

5   Conclusion

In this paper, aiming at the task assignment problem of multiple USVs in attack strategy, the DAMGWO algorithm, adding the DAM to improve GWO algorithm, is introduced. During the process of wolf initialization and algorithm executing, the initial wolves are generated by DAM according to practical constraints. It ensures the diversity of the grey wolves, and avoids exhaustion and plunging into the local optimum. The simulation results show that this improved GWO algorithm can be applied to the field of multiple USVs in attack strategy, and its efficiency is better than traditional GWO and PSO algorithm.

Acknowledgments. This paper is supported by the National Natural Science Foundation of China (Grant No. 61625304), by the Science and Technology Commission of Shanghai Municipality (16511102400) and by the Innovation Program of Shanghai Municipal Education Commission under Grant No. 14YZ024.

References

1. Zhang, B., Wang, L., Li, Y.: Development trend of unmanned surface vehicle. Sci. Technol. Vis. **19**, 301 (2016)
2. Chen, H.Y., Zhang, Y.: An overview of research on military unmanned ground vehicles. Acta Armamentarii **35**, 1696–1706 (2014)
3. Bo, W.U., Wen, Y., Bei, W.U., Zhou, S., Xiao, C., Navigation, S.O.: Review and expectation on collision avoidance method of unmanned surface vessel. J. Wuhan Univ. Technol. **40**(3), 456–461 (2016)
4. Steimle, E.T., Murphy, R.R., Lindemuth, M., Hall, M.L.: Unmanned marine vehicle use at Hurricanes Wilma and Ike, pp. 1–6. IEEE (2009)

5. Yang, N., Guo, D.L.: Solving polynomial equation roots based on wolves algorithm. Sci. Technol. Vis. **15**, 35–36 (2016)
6. Wang, X.F., Wang, B.: Techniques for threat assessment based on intuitionistic fuzzy theory and plan recognition. Comput. Sci. **37**(5), 175–177 (2010)
7. Zhang, B., Kang, F.J., Bing, S.U., Collage, M.: Simulation of allied attack mission planning model of multi- unmanned surface vessels. Comput. Simul. **32**(4), 349–354 (2015)
8. Xhafa, F., Sun, J., Barolli, A., Biberaj, A., Barolli, L.: Genetic algorithms for satellite scheduling problems. Mob. Inf. Syst. **8**(4), 351–377 (2012)
9. Ceriotti, M., Vasile, M.: Automated multigravity assist trajectory planning with a modified ant colony algorithm. J. Aerosp. Comput. Inf. Commun. **7**(9), 261–293 (2010)
10. El-Menshawy, M., Bentahar, J., El Kholy, W., Dssouli, R.: Verifying conformance of multi-agent commitment-based protocols. Expert Syst. Appl. **40**(1), 122–138 (2013)
11. Pu, J., Wu, X., Guo, Y., Xie, S., Pu, H., Peng, Y.: The combat of unmanned surface vehicles based on wolves attack. Intell. Softw. Methodol. Tools Techn. **297**, 794–805 (2017)
12. Yu, J.P., Zhou, X.M., Chen, M.: Research on representative algorithms of swarm intelligence. Jisuanji Gongcheng yu Yingyong (Comput. Eng. Appl.) **46**(25), 1–4 (2010)
13. Li, J., Sun, Y.: Research on cluster attack mission planning of USVs based on DAMPSO algorithm. Comput. Eng. Appl. **49**(20), 1–4 (2013)
14. Mirjalili, S., Mirjalili, S.M., Lewis, A.: Grey wolf optimizer. Adv. Eng. Softw. **69**, 46–61 (2014)
15. Adhau, S., Mittal, M.L., Mittal, A.: A multi-agent system for distributed multi-project scheduling: an auction-based negotiation approach. Eng. Appl. Artif. Intell. **25**(8), 1738–1751 (2012)
16. Long, W., Zhao, D., Xu, S.: Improved grey wolf optimization algorithm for constrained optimization problem. J. Comput. Appl. **35**(9), 2590–2595 (2015)

A Study of Multi-label Event Types Recognition on Chinese Financial Texts

Shunxin Luo, Yinglin Wang[✉], Xue Feng, and Zhenda Hu

School of Information Management and Engineering,
Shanghai University of Finance and Economics, Shanghai 200433, China
wang.yinglin@shufe.edu.cn

Abstract. Event extraction is a technique that automatically extracts key event elements from large-scale texts. In classic event extraction process, recognizing event types is earlier than extracting argument roles in the whole event extraction task. But in practice, multiple-label problems are often encountered (that is, one event sentence corresponds to multiple event types). In order to solve this problem, this paper introduces Binary-Relevance, Classifier-Chain, MLkNN and many other multi-label classification strategies from the perspective of problem transformation and algorithm adaptation, trying to find the best classification method to adapt to our Chinese financial corpus. The experimental results show that the Adaboost method based on single-layer decision tree with Classifier-Chain is the best strategy for the task of recognizing event types in this paper. The micro-F1 score and average-precision value of this strategy are 8.91% and 12.46% higher than the baseline strategy (Binary-Relevance + SVM) respectively. At the same time, this method achieves the lowest value on the three indicators of Hamming-Loss, Coverage-Error and Ranking-Loss. In addition, the results also show: (1) Classifier-Chain strategy is better than Binary-Relevance strategy when the classifiers are the same; (2) Under the same problem transformation strategy, the Adaboost method performs best, followed by KNN and the worst case is SVM; (3) If only single classifier is allowed, the MLkNN strategy based on algorithm adaptation is better than other strategies based on problem transformation.

Keywords: Event types recognition · Multi-label classification
Financial text analysis

1 Introduction

As the pre-step of the feature extraction task, whether the event type can be correctly identified affects the performance of the argument extraction system directly. In essence, recognizing event types can be regarded as a text categorization task which classifies the event description sentences into the event class. Considering that there is a certain mapping relationship between trigger words and event types, the previous studies often used trigger word extraction as a pre-order task for event type recognition. For example, Ding [1] regards the verb involved in both subject-verb relationship and the verb-object relationship in a sentence as the event trigger according to syntactic analysis, and the event types which the sentence statement belongs to are judged by the

© Springer Nature Switzerland AG 2018
S. Wrycza and J. Maślankowski (Eds.): SIGSAND/PLAIS 2018, LNBIP 333, pp. 146–158, 2018.
https://doi.org/10.1007/978-3-030-00060-8_11

context of the words around the trigger. However, this method has certain flaws. For instance, triggers must be selected beforehand from words of sentences that may contain many non-triggers. Thus, it may introduce too many counterexamples. This method may also ignore triggers that are not included in the subject-verb-object relationship or some noun triggers, while this situation appears in our corpus. Similar methods were also adopted by other researchers, such as Ahn et al. [2], Zhao et al. [3] and Qin et al. [4]. They presented methods that combines event trigger expansion and binary classifiers in the event type recognition task.

Considering these flaws, some researchers introduced the other event type recognition methods without triggers. Tan [5] applied binary relevance strategy with a local feature selection method, which can effectively reduce dimension of classification word features and achieve 83.5% F1-score (the best result in his research) in ACE2005 corpus. Xu [6] assigned each candidate event instance with corresponding event type by multi-classification strategy. In Xu's research, classification features are presented as an aggregation with words that appear in sentences and semantics. These studies indicated that using multi-classification without triggers to recognize event types can achieve good experimental results.

Besides of that, an event may belong to multi-types, which is called the multi-label problem and it refers to a situation that one sample corresponds to multiple categories [7]. Take the research field of this article as an example. "Huasheng Jiangquan Company has suffered a loss in 2016 and faced a relatively large repayment pressure. It was included in the credit rating watching list by Pengyuan Credit." This sentence can be classified as financial affairs and can also be regarded as a rating event. If we only classify it into a single category, it will be bound to lose part of the information and will cause incorrect extraction of event elements.

Because traditional classification algorithms are unable to solve multi-label problems directly, people put forward solutions from the perspectives of problem transformation and algorithm adaptation. The so-called problem transformation strategy refers to transforming multi-labels into single labels to adapt to existing classification algorithms. The classical methods include Binary-Relevance, Label-Power-Set, Classifier-Chain, and Calibrated-Label-Ranking. The idea of adaptation strategy is just the opposite, which can directly handle multi-label problems by changing the existing classifiers. The most famous of these methods is the improved MLkNN model based on KNN. By maximizing the posterior probability and the Bayesian criterion, the final label types of multi-label samples can be obtained [8].

In this paper we treat event recognition as a text classification task and introduce it into the study of Chinese financial text for the first time. In the study, event types are identified without using trigger words. Moreover, considering multi-label problems which appear in our task, we introduce and compare several multi-label classification strategies from the two perspectives of problem transformation and algorithm adaptation, such as Binary-Relevance, Classifier-Chain and MLkNN in the event type identification module in order to find the best strategy in the task.

2 Method Description

2.1 Methods Based on Problem Transformation

In this section, we first describe the three multi-label methods based on problem transformation. We assume that data set D has a total of m kinds of label categories, and each instance corresponds to at least one or more labels.

Binary-Relevance: The idea of Binary-Relevance strategy is similar to One-vs-all in multi-category classification. It divides the original data set into m (the same number of label categories) binary data sets and each data set D_j contains the whole original data instances. For each D_j, we classify the instances belonging to the j-th class as positive, and the rest of the instances as negative. Then the original multi-label classification problem is translated into m dichotomous problems. The Binary-Relevance strategy is one of the first-order strategies, and it has the advantage of low computational complexity, which is more applicable to the cases with fewer label categories. But its shortcoming is also obvious, that is, it only considers the labels as independent individuals, ignoring the correlation among them.

Assuming that the current corpus aggregate has 5 sentences (S_1–S_5) corresponding to the three event types (y_1–y_3), the strategy of Binary-Relevance can be illustrated in Fig. 1.

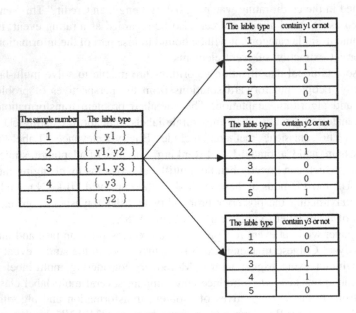

Fig. 1. An example of Binary-Relevance strategy

Classifier-Chain: The Classifier-Chain strategy can be considered as an extension of the Binary-Relevance strategy. This strategy also divides the raw data set into m binary data sets. But the difference is that it orderly sorts multiple base classifiers (or divided data sets) into one classifier chain, and then performs classification experiments in sequence. In the process, the results of the previous stage will be passed as a part of input to the following stage. Since the tag information can be passed between multiple stages, the Classifier-Chain strategy takes into account the correlation between the tags, and it can compensate for the defects of Binary-Relevance strategy in this respect [9].

Label-Power-Set: Different from the Binary-Relevance strategy, Label-Power-Set strategy regards each multi-label set as a new label. Thus, the correlation between labels is considered but it makes the labels more complicated, and the new labels may only has very few instances, and result in data imbalance.

After a multi-label classification problem is converted to single label classification problems through a problem transformation strategy, traditional classifiers can be used. In our experiments, we compare support vector machine (SVM), K-Nearest Neighbor (KNN), and Adaboost, which are commonly used in classification tasks. Support Vector Machine (SVM) is to obtain a segmented hyperplane in a multidimensional space, and then classify instances into different sides of the hyperplane. KNN belongs to a lazy learning algorithm, which counts the number of categories of the K-nearest neighbors in the feature space of the instances, and then uses the largest number of categories as the prediction category of the instance. Unlike the above two classifiers, Adaboost is an ensemble learning method that implements model adaptive enhancement. It adopts the idea of iteration and adds a new base classifier at each iteration. In addition, it increases the weight of the instance that was previously misclassified and reduces the one of correctly categorized instance. Through numbers of iterations, the final strong classifier can be obtained. In this paper a single-level decision tree is selected as the base classifier for Adaboost.

2.2 Method Based on Algorithm Transformation

Of the algorithm-based adaptation strategy, the MLkNN method is used in this paper. Based on the traditional K-nearest neighbors algorithm, this method counts the class labels of the latest K instances from the instances, and then infers the new instance's label set by maximizing the posterior probability.

Assuming that the multi-label test set is $\{(x_i, Y_i)|1 \leq i \leq m\}$, where x_i is the instance' feature and Y_i is the actual label set corresponding to x_i. For an unpredicted instance $\bar{x}$, the set of K instances around it is called $N(\bar{x})$, then for any class label $y_j(1 \leq j \leq n)$ in the label space, we can count the number of instance which contain y_j in $N(\bar{x})$. Here it is denoted as C_j:

$$C_j = \sum_{(x, Y) \in N(\bar{x})} [\![y_j \in Y]\!] \tag{1}$$

Based on this, we use the symbol H_j to indicate the event that the unpredicted instance $\bar{x}$ has the class label y_j. Correspondingly, $P(H_j|C_j)$ represents the posterior probability that the event H_j holds when there are C_j samples with the class label y_j in the $N(\bar{x})$; $P(\overline{H_j}|C_j)$ represents the posterior probability that the event H_j does not holds under the same condition. Then classifier I can predict the classes the instance $\bar{x}$ belongs to using the following formula:

$$I(\bar{x}) = \{y_j | P(H_j|C_j)/P(\overline{H_j}|C_j) > 0.5\}(1 \leq j \leq n) \tag{2}$$

According to Bayes' theorem, the above equation can also be expressed as

$$I(\bar{x}) = \{y_j | P(C_j|H_j) \times P(H_j)/P(C_j|\overline{H_j}) \times P(\overline{H_j}) > 0.5\}(1 \leq j \leq n) \tag{3}$$

$P(C_j|H_j)$ indicates the conditional probability that when the event H_j holds, there are C_j samples with the class label y_j in the $N(\bar{x})$; $P(H_j)$ represents the prior probability that H_j holds; $P(C_j|\overline{H_j})$ and $P(\overline{H_j})$ can be interpreted in a similar way. Thus, through statistics of the training examples s, the class labels of $\bar{x}$ can be predicted.

2.3 Vector Space Model and Feature Weights

For natural language processing tasks, the texts need to be firstly translated into a machine-readable form. In this regard, researchers have proposed a variety of representational models that deal with textual data, such as Boolean models, vector space models, probabilistic models, latent semantic models, and word vector models which have received much attention in recent years. Among them, the vector space model (VSM) is the most widely used. It was proposed by Salton (1975) and it can map a single document into a vector in a high-dimensional space, thereby realizing the vectorization of the document. The VSM model assumes that the text set D has a total of n features $t_1, t_2, t_3 \ldots t_n$, which can be used to represent any text d in D, hence, text d in D can be represented as the following feature vector:

$$V(d) = \{(t_1, w_1), (t_2, w_2), (t_3, w_3) \ldots (t_n, w_n)\} \tag{4}$$

Among them, t_i denotes the features of the text, which can be language units of different forms such as words, phrases and sentences. $w_i (i = 1, 2, 3 \ldots n)$ represents the feature weight corresponding to the feature item t_i.

The above vector space model (VSM) expression contains two parts of feature t and feature weight w. In practice, we can assign values to w according to different calculation methods. Common feature weights include Boolean weight, word frequency weight, IDF weight, and TFIDF weight. In this paper, we consider word frequency as feature weight. It indicates the number of occurrences of the feature item in the document. This method shows that if the frequency of a feature item in a document is higher, the feature item is more representative and its formula is as follows:

$$w_{ij} = c_{ij} \tag{5}$$

c_{ij} denotes the frequency of occurrence of feature item i in document j, and w_{ij} denotes the feature weight value of feature i in the corresponding document j.

2.4 Multi-labeled Text Feature Selection Method

As we mentioned above, the vector space model implements the mapping of text data to a high-dimensional vector space. However, for large-scale documents, with the increase of the number of words, the space dimension will expand rapidly, and then bring about the problem of "dimensional disasters" [10]. In this regard, people propose a series of feature selection methods. Through certain screening criteria, those redundant features that cannot characterize the feature of the document are eliminated, so as to improve the performance with a feature space of reduced dimension.

Several feature selection methods were developed, and commonly used ones include Chi-square, Mutual Information, Information Gain, and Expected Cross Entropy. Past practice shows that Chi-square and information gain can often achieve good experimental results. Tan [5] selected Chi-square and information gain in a classification task. The experimental results showed that there was no significant difference in classification performance between the two methods. Therefore, in this paper we directly select the chi-square strategy as the feature selection method in the experiment.

The χ^2 statistic is an important concept in statistics. It first assumes that the two variables are independent, and then compares the deviation between the actual measurement and the theoretical measurement to determine whether the null hypothesis is true. If the value of χ^2 is large enough, the original hypothesis is rejected and the alternative hypothesis is accepted. The correlation between the two variables is considered to be strong. If the χ^2 value is small enough, the deviation is considered to be a natural sample error and the original hypothesis that the two variables are independent should be accepted. The following formula shows a method for calculating the χ^2 value between the term feature t_i and the class l_j, where t_i denotes the feature, $P(t_i, l_j)$ denotes the frequency of the term t_i that relates to the l_j class in the data set, $P(t_i)$ denotes the term t_i's frequency of occurrence in the entire data set. The larger the χ^2 value, the stronger the correlation between this feature and the l_j class.

$$\chi^2(t_i, l_j) = \frac{n\left[P(t_i, l_j)P(\overline{t_i}, \overline{l_j}) - P(t_i, \overline{l_j})P(\overline{t_i}, l_j)\right]^2}{P(t_i)P(\overline{t_i})P(l_j)P(\overline{l_j})} \tag{6}$$

For multi-labeled event texts, it is assumed that there are n entries and m event types in the data set. In the actual computation of the chi-squared value, for each term's feature t_i, we can find m chi-squared values for all event classes $\chi^2(t_i, l_1), \chi^2(t_i, l_2) \ldots \chi^2(t_i, l_m)$. This article takes the maximum value of all categories of chi-squared values of t_i as the score $\chi^2(t_i)$ of this term. For the n term features in the data set, a total of n scored items can be calculated, and the scored items are sorted in descending order. Additionally, the first k features are then selected to compose the feature subset. In the following experiments, feature subset of this kind will be used.

$$\chi^2(t_i) = \max\left\{\chi^2(t_i, l_1), \chi^2(t_i, l_2) \ldots \chi^2(t_i, l_m)\right\} \tag{7}$$

In addition, for the Adaboost algorithm based on a single decision tree, it will select one feature from all features to achieve the optimal value of the criterion function in each weak classifier training process, so Adaboost itself has good features, with no need to use the chi-squared strategy mentioned above.

3   Case Study

3.1   Corpus Description

Our special task is to extract interesting financial events from financial research report of companies. We sum up seven types of financial event, namely rating, assets, securities issue, personnel, financial affairs, macro policy and index. For the 4,913 candidate event sentences obtained, we manually annotated the event types. Firstly, the three students marked the first 500 sentences together, and then discussed their respective labels. Based on unified rules, the three students annotated the remaining 4,413 sentences by the division of labor. Table 1 shows the statistics of the annotated sentences. It should be noted that due to the multi-label problem (one sentence may correspond to multiple event types), the total number of event sentences summed by category is greater than 4,913.

Table 1. The statistics of the event types in the corpus

Event type	Numbers
Rating	1,272
Securities issue	1,033
Personnel	342
Assets	670
Financial affairs	327
Macro policy	406
Index	438
Other	600
Total	4,913

3.2   Experimental Design

In this section, we mainly explore the differences of results in the task of event type recognition of sentences in financial research reports, under different multi label classification strategies. We select three multi label classification strategies - MLkNN, Binary-Relevance and Classifier-Chain. For the last two strategies, we carry out the contrastive experiment with KNN, SVM and Adaboost respectively. We have a total of seven groups of experimental models and they are (1) Binary-Relevance + SVM; (2) Classifier-Chain + SVM; (3) Binary-Relevance + KNN; (4) Classifier-Chain + KNN; (5) Binary-Relevance + Adaboost; (6) Classifier-Chain + Adaboost; (7) MLkNN.

Before the experiment, firstly we use the VSM model to process the event sentences and select the word frequency as the feature weight. After that, the feature selection method introduced in the second section is used. For the data of experiment, the corpus is randomly divided into training set, validation set and test set according to the proportion of 3:1:1.

3.3 Evaluation Criterion

Because of the particularity of multi-label problem, the common single classification evaluation criteria may not be directly applicable. Therefore, in this part, we introduce 5 evaluation indicators to evaluate the effect of multi-label classification methods.

Among the 5 indicators, four are indicators based on examples and one is based on categories.

Before further explanation, we first make some notations. Here, the multi-labeled test set is denoted as $\{(x_i, Y_i)|1 \le i \le m\}$, where x_i is the feature of an instance and Y_i is the corresponding real label set. Y_i is a subset of all class space $L(Y_i \subseteq L)$. The label set predicted by a classifier is represented as $I(x_i)$. At the same time, the classifier corresponds to a real-valued function f. $f(x_i, y)$ can be regarded as the confidence level of an instance x_i corresponding to a label y.

The four indicators based on the samples used in this paper includes Hamming-Loss, Ranking-Loss, Coverage-Error and Average precision. They are as follows:

1. *Hamming-Loss*
   This indicator is used to measure the difference between the predicted labels and actual labels. The smaller the value is, the closer to the real situation the predicted labels are. The calculation formula is expressed as follows:

$$\text{Hamming-}Loss = \frac{1}{m} \sum_{i=1}^{m} \frac{|I(x_i) \, \Delta \, Y_i|}{n} \tag{8}$$

   where Δ represents the number of differences between the set $I(x_i)$ and Y_i.

2. *Ranking-Loss*
   This indicator calculates the possibility of unrelated labels having higher ranking order than related labels. Similarly, the smaller the value of the index is, the better the performance of the algorithm is. The calculation formula is expressed as follows:

$$Ranking\text{-}Loss = \frac{1}{m} \sum_{i=1}^{m} \frac{1}{|Y_i||\overline{Y_i}|} |\{(y, \bar{y})|f(x_i, y) \le f(x_i, \bar{y}), (y, \bar{y}) \in Y_i \times \overline{Y_i}\}| \tag{9}$$

   where $\overline{Y_i}$ represents the complement of Y_i in the label space L.

3. *Coverage-Error*
   This indicator represents the search depth needed to cover all relevant labels for a label sequence. Similar to the above indicators, the smaller the index value is, the better the performance of the model is.

$$Coverage\text{-}Error = \frac{1}{m}\sum_{i=1}^{m} max_{y \in Y_i} rank(x_i, y) - 1 \tag{10}$$

where *rank* represents the sorting function corresponding to *f*.

4. *Average-Precision*

The indicator calculates the probability that the label before a correct label in the predicted label sequence is still correct. r. Therefore, the bigger the index value is, the better the classification effect is.

$$Average\text{-}Precision = \frac{1}{m}\sum_{i=1}^{m} \frac{1}{|Y_i|} \sum_{y \in Y_i} \frac{|\{y' | rank(x, y') \le rank(x_i, y), y' \in Y_i\}|}{rank(x_i, y)}$$

$$\tag{11}$$

5. *The F1 score of Macro-Averaged*

$$Macro\_F1 = \frac{2 \times Macro\_Precision \times Macro\_Recall}{Macro\_Precision + Macro\_Recall} \tag{12}$$

The calculation of the F1 score of Macro-Averaged-value depends on *Macro_Precision* and *Macro_Recall*. Their formulas are shown as follows respectively:

$$Macro\_Precision = \frac{1}{n}\sum_{i=1}^{n} P_i \tag{13}$$

$$Macro\_Recall = \frac{1}{n}\sum_{i=1}^{n} R_i \tag{14}$$

3.4   Experimental Results and Analysis

3.4.1   Determination of the Optimal Feature Dimension

Determining the appropriate feature dimension is very important for the final classi-fication experiment. In this part, Micro-F1, Hamming-Loss and Ranking-Loss are used as measurement criterion to examine the experimental results of strategies (1)–(4) and (7) (refer to Sect. 3.2) on the validation set with different *k* values. The range of *k* is from 10 to 550.

The experimental results are shown in the following diagrams, in which the hori-zontal axis represents the feature dimension, the left vertical axis is the indicator value of Micro-F1 and the right vertical axis is the indicator value of Hamming-Loss and Ranking-Loss. The near neighbor number in MLkNN and KNN is set to 10.

From the graphs from Figs. 2, 3, 4, 5 and 6, it can be found that with the increase of the feature dimension, Micro-F1 first increases and then drops while Hamming-Loss and Ranking-Loss first descend and then increase. It shows that the effective features obtained via the Chi-square selection method are ranked sequentially at the feature set. The more effective the feature, the smaller the ordinal number of the feature in the set. With the increasing number of features, some redundant features are introduced into the model, resulting in a gradual decline in classification performance. According to the turn point, for strategies (1)–(4), the number of the feature dimension is between 70 and 130. And Micro-F1 gets the maximum value while Hamming-Loss and Ranking-Loss get the minimum value. In contrast, the performance of MLkNN strategy (7) is more stable, and it can achieve better experimental results when the number of dimensions is in the range of 70 and 200. In order to unify the condition, we set the experimental feature dimensions of the five strategies (1)–(4) and (7) to 100.

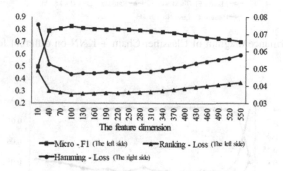

Fig. 2. The performance diagram of Binary-Relevance + SVM on different feature dimensions

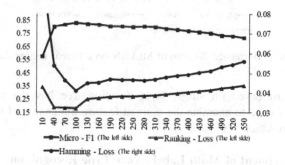

Fig. 3. The performance diagram of Classifier-Chain + SVM on different feature dimensions

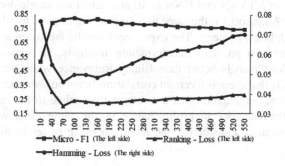

Fig. 4. The performance diagram of Binary-Relevance + KNN on different feature dimensions

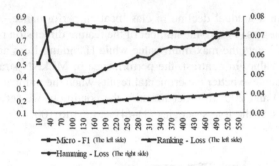

Fig. 5. The performance diagram of Classifier-Chain + KNN on different feature dimensions

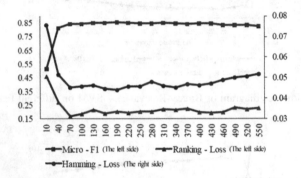

Fig. 6. The performance diagram of MLkNN on different feature dimensions

Since Adaboost based on single layer decision tree has the feature selection function, we don't need to examine the optimal feature dimension of the strategies (5) and (6) on the validation set.

3.4.2   The Experiment of Multi Label Event Type Recognition

Based on the above feature dimension selection schemes, we test the performance of strategies (1)–(7) on the test set respectively. In terms of parameters, we set the nearest neighbor number of MLkNN and KNN to 10 and select the single decision tree as the base classifier for Adaboost. At the same time, we set the number to 50 and the learning rate to 1.0 for Adaboost strategy. The experiment results are listed at Table 2.

Comparing the two problem transformation methods, the classification strategy based on Classifier-Chain is better than Binary-Relevance when the classifier is the same. The reason is that there is a certain correlation between the types of events in this paper. For example, the situation that a sentence may relate to the category of rating adjustment and the category of profit change in the same time, which makes the Classifier-Chain strategy considering the label relevance better in the corpus of this research.

Table 2. The experimental results of multi label event type recognition

	BR + SVM	CC + SVM	BR + KNN	CC + KNN	BR + ADA	CC + ADA	MLkNN
Hamming-Loss (1)	0.050	0.043	0.045	0.046	0.039	**0.036**	0.042
Ranking-Loss (2)	0.283	0.200	0.202	0.166	0.163	**0.119**	0.160
Coverage-Error (3)	3.078	2.499	2.518	2.269	2.226	**1.911**	2.213
Average-Precision (4)	63.89%	68.80%	68.33%	70.18%	73.54%	**76.35%**	72.17%
Micro-F1 (5)	79.75%	80.69%	80.80%	83.24%	87.72%	**88.66%**	85.38%

Comparing the three classifiers under the same problem transformation strategy, the Adaboost method based on single layer decision tree is the best, followed by KNN and the worst is SVM, which shows that the adaptive enhancement ensemble classifier is superior to a single classifier. In addition, the classification performance of KNN is slightly better than that of SVM. The reason may be that KNN can predict the category of the unknown instance by the limited surrounding examples instead of judging the class according to the partition of class spaces, so the effect of the KNN classification may be better for the case where overlapping of the class spaces exists.

In overall, the Adaboost method based on Classifier-Chain (Classifier-Chain + Adaboost) performs best on the test set, whose Micro-F1 and Average-Precision values are 8.91% and 12.46% better respectively than Binary-Relevance + SVM, CC + ADA gets the lowest values on Hamming-Loss, Coverage-Error and Ranking-Loss. If we only consider the situation with single classifier, we found that MLkNN based on algorithm adaptation strategy is better than other methods based on the problem transformation strategy. It is similar to the results obtained by Xu [11] on other open datasets. In summary, the performance of the classification methods used in our study is listed as follows:

$$CC + ADA > BR + ADA \approx MLkNN > CC + KNN \geq CC + SVM$$
$$\approx BR + KNN > BR + SVM$$

4 Conclusion

This paper studies the methods for identifying the event types of sentences in financial text. In this research, the identification of event category is treated as a text classification task. The candidate event sentences are firstly vectorized using the VSM model, and then appropriate feature selection strategies of are used to reduce the dimension of features, and subsequent classification experiments are carried out. Due to the multi-label problem in the corpus of this paper, the traditional classification methods are difficult to be used directly. Therefore, starting from the problem transformation and algorithm adaptation, we introduce and study multiple multi-label classification models and try to find the best classification strategy via experiments.

The experiments show the following conclusions: (1) Comparing two kinds of problem transformation methods in the case of the same classifier, Classifier-Chain strategy is slightly better than Binary-Relevance strategy; (2) Comparing two classifiers under the same problem transformation strategy, the Adaboost method based on single-level decision tree performs best, followed by KNN, and the worst is SVM; (3) If a single classifier is used, the MLkNN method based on algorithm adaptation is better than other problem-transformation-based methods. In general, the Adaboost method with single-layer decision tree based on Classifier-Chain is the best solution to the task of identifying event types of Chinese financial text in this paper.

In the future, we will try to explore semi-supervised ways to combine the unlabeled data and improve the performance further.

Acknowledgement. We thank the National Natural Science Foundation of China (No. 61375053) for financial support.

References

1. Ding, X.: Research on sentence level Chinese event extraction. Master dissertation, Harbin Institute of Technology (2011). (in Chinese)
2. Ahn, D.: The stages of event extraction. In: Proceedings of the Workshop on Annotating and Reasoning about Time and Events, Sydney, Australia, 23 July 2006. Association for Computational Linguistics (2006)
3. Zhao, Y.Y., Qin, B., Che, W.X., Liu, T.: Research on Chinese event extraction. J. Chin. Inf. Process. **22**(1), 3–8 (2008). (in Chinese)
4. Qin, B., Zhao, Y.Y., Ding, X., et al.: Event type recognition based on trigger expansion. J. Tsinghua Univ. (Sci. Technol.) **15**(3), 251–258 (2010)
5. Tan, H.Y.: Research on Chinese event extraction. Doctoral dissertation, Harbin Institute of Technology (2008). (in Chinese)
6. Xu, H.L., Chen, J.X., Zhou, C.L., et al.: Research on event type identification for Chinese event extraction. Mind Comput. **4**(1), 34–44 (2010). (in Chinese)
7. Tsoumakas, G., Katakis, I., Taniar, D.: Multi-label classification: an overview. Int. J. Data Warehous. Min. **3**(3), 1–13 (2007)
8. Zhang, M.L., Zhou, Z.H.: ML-KNN: a lazy learning approach to multi-label learning. Pattern Recogn. **40**(7), 2038–2048 (2007)
9. Read, J., Pfahringe, B., Holmes, G., et al.: Classifier chains for multi-label classification. Mach. Learn. **85**, 333–359 (2011)
10. He, J.Y., Chen, R., Xu, M., et al.: Algorithm for Chinese text categorization based on class feature vector representation. Appl. Res. Comput. **25**(2), 337–341 (2008)
11. Xu, Z.G.: A comparative study of multi-label classification approaches. http://lamda.nju.edu.cn/huangsj/dm11/files/xuzg.pdf

Healthcare IT

Implementation of ICT for Active and Healthy Ageing: Comparing Value-Based Objectives Between Polish and Swedish Seniors

Ella Kolkowska[1]([⊠]), Ewa Soja[2], and Piotr Soja[3]

[1] Center for Empirical Research in Information Systems (CERIS),
Örebro University School of Business, Örebro, Sweden
Ella.Kolkowska@oru.se
[2] Department of Demography, Cracow University of Economics,
Kraków, Poland
Ewa.Soja@uek.krakow.pl
[3] Department of Computer Science, Cracow University of Economics,
Kraków, Poland
Piotr.Soja@uek.krakow.pl

Abstract. Active and healthy ageing strategies are proposed in many European countries to address the challenges generated by ageing of the populations. Information and communication technology (ICT) plays an important role in the implementation of these strategies. By applying Value-focused thinking approach, this qualitative study investigates what objectives are important for successful implementation of ICT for active and healthy ageing according to older people in Sweden and Poland. The study shows that there are both differences and similarities between the objectives identified in these two countries that may have significant implications for development (analysis and design) and implementation of ICT solutions for active and healthy ageing.

Keywords: ICT · Active and healthy ageing · Value-based thinking
Poland · Sweden

1 Introduction

An increase in the share of older people in the population is a typical phenomenon of developed countries [3]. Such a process has a profound influence on the whole economy and society. In particular, many of the concerns involve projected increase of age-related expenditures including pensions, health care, and long-term care [1, 4]. To counteract the effects of an ageing population various strategies are proposed and are now being developed [6], for example, active ageing [21], healthy ageing [17] or successful ageing [23]. These strategies should enable and promote longer working lives, ensure that private family transfers are integrated into old-age security systems where possible, promote well-being and enable healthy active living to reduce chronic illness and health care costs, and support active contributory life for as long as possible [8].

© Springer Nature Switzerland AG 2018
S. Wrycza and J. Maślankowski (Eds.): SIGSAND/PLAIS 2018, LNBIP 333, pp. 161–173, 2018.
https://doi.org/10.1007/978-3-030-00060-8_12

An important role in the implementation of these strategies is played by information and communication technology (ICT) [9, 20]. The role of ICT is particularly acknowledged by decision-makers in the field of healthcare and long-term care (e.g., [18, 22]). However, on the one hand, the possibilities of using ICT in this domain can depend on wide-ranging contextual factors such socio-economic and demographic factors (e.g., level of economic development, population ageing level, solutions in the field of social and health policy, and digital development level) [19, 22]. On the other hand, for ICT solutions to be used by older people, it is necessary to take into account their personal needs in this area [16]. In this respect, Kapadia et al. [10] emphasize that involving older people during the design process is critical for developing technologies that meet the needs of the elderly.

Against this backdrop, we argue that older people together with other stakeholders should be involved in developing strategies for implementation of ICT for active and healthy ageing. The study aims to identify objectives that older people in Poland and Sweden consider as especially important in this context.

Sweden and Poland are very different regarding demographic and socio-economic aspects [13, 15] hence it is interesting to see if the identified objectives differ between these two countries. Our study seeks to answer the following research questions: *(1) What objectives are important to include in strategies for implementation of ICT for active and healthy ageing according to older people in Sweden and Poland?, (2) What are the differences and similarities between the identified objectives in these two countries?*

To identify the objectives, the current study applies Value-focused thinking (VFT) approach. VFT was put forward by Keeney [11] to improve decision-making in a specific context by grounding strategic decisions in values identified in that context. VFT can be used for formulating objectives based on values held by different stakeholders [7]. Nevertheless, in this exploratory study, we focus only on older people.

The paper is organized as follows. In the next section, we present research background. Next, we describe our research method, which is followed by the presentation of results. We then discuss our findings, explain implications, and close the study with concluding remarks.

2   Background

On the basis of Eurostat database [5] we noted several differences between Sweden and Poland concerning demographic and socio-economic aspects. First, percentage of older people (65+) in the population is 19.6% in Sweden and 15.6% in Poland. Second, life expectancy at birth is higher in Sweden than in Poland (Sweden 83.7 women and 80.1 men; Poland 81.9 women and 73.9 men). Third, number of healthy life years after age 65 is significantly higher for Swedes than Poles (in Sweden 16.8 for women and 15.7 for men; in Poland 8.4 for women and 7.6 for men). Fourth, a higher number of older people in Sweden live in single households (40% in Sweden, 30% in Poland). Fifth, both countries differ in economic development level: Poland's GDP per capita in PPS (Purchasing Power Standards) for 2016 was 69 while Sweden's was almost twice as high: 124 in the same year [5]. Finally, as illustrated in Fig. 1, Sweden belongs to countries where the Internet use by people aged 55–74 is the highest, while Poland is located among countries with the lowest percentage of such people in the whole population.

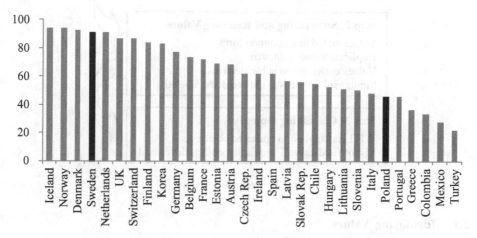

Fig. 1. Internet use by individuals aged 55–74 in 2017 (%) (https://stats.oecd.org/)

Furthermore, Swedish and Polish elderly care systems are organized differently [13]. The Swedish system is built on state responsibility model with strong emphasis on redistribution, social inclusion and universality of public services. In Poland, only some of the care needs are satisfied by the government, while other services are rendered by families and private service organizations (e.g., [2, 13]). Swedish Government puts significant resources on digitalization of health care and social care. In Poland, innovative digital solutions used in elderly care are still rare [19]. Prior studies in Poland and Sweden revealed that seniors in both countries generally have positive attitudes towards ICT solutions, but they differ with respect to requirements and expectations towards technologies supporting independent living [15].

3   Research Method

The study was conducted in line with VFT [11]. According to VFT, the process of identifying value-based objectives starts with interviews with the concerned people. Since our exploratory study focuses on objectives important for older people, we conducted in-depth interviews with 15 older people in Sweden and 15 seniors in Poland. To ensure a relatively good representativeness and diversity of samples for both countries, we have chosen respondents of both genders (7 men and 8 women), aged 65–85, with different places of residence and varied health status. Following VFT [11] the value-based objectives were derived from the interviews in three steps (Fig. 2).

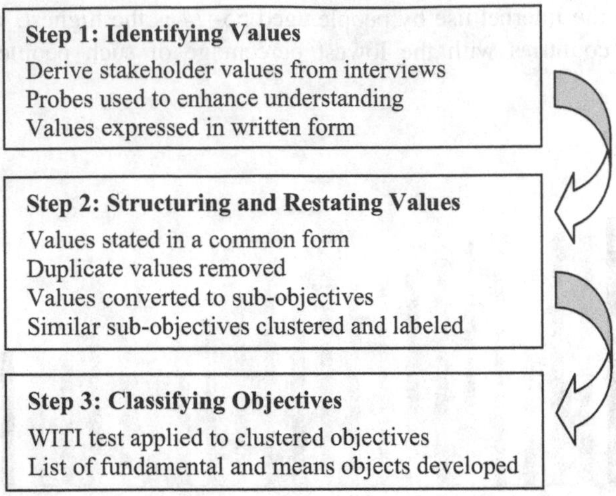

Step 1: Identifying Values

Derive stakeholder values from interviews
Probes used to enhance understanding
Values expressed in written form

Step 2: Structuring and Restating Values

Values stated in a common form
Duplicate values removed
Values converted to sub-objectives
Similar sub-objectives clustered and labeled

Step 3: Classifying Objectives

WITI test applied to clustered objectives
List of fundamental and means objects developed

Fig. 2. Research approach

3.1   Identifying Values

The interviews were conducted following Keeney [12] and focused on the respondents' wishes and needs regarding implementation and use of ICT for active and healthy ageing. At the beginning of each interview, we clarified the purpose and scope of the interview. Since most of the people have difficulty with expressing values, we used suggested by Keeney [11] words such as trade-offs, consequences, impacts, concerns, fair and balance, to trigger questions which would make implicit values more explicit. To find values in the collected material, we focused on statements expressing problems, consequences, better or worse alternatives, or goals. In line with Keeney's suggestion [12, p. 33], we adopted the definition of values as 'principles for evaluating the desirability of any possible alternatives or consequences' The analysis resulted in a list of statements that were numbered, written as values, and input into a database.

3.2   Structuring and Restarting Values

First, identified values were structured in a common form in order to eliminate all duplicates. Second, each value was converted into an objective, since according to VFT, each value is anchored in an objective [7] and can be easily translated into that objective. All objectives were then analyzed to identify those objectives that deal with similar issues. By the end of this step, all objectives dealing with similar issues were categorized into categories, which were labeled. In the first step, the objectives identified in the Polish and Swedish contexts were categorized separately. Then, the categories were discussed in a group of researchers involved in this study and changed if needed.

3.3 Classifying Objectives

The categories were then organized into two groups in order to understand how the objectives are related to each other in the given decision context. In this study, the decision context is the implementation of ICT for active and healthy ageing. The categories of objectives were organized into two categories: 'fundamental' and 'means' objectives. Fundamental objectives are essential objectives in a given decision context, whilst means objectives help to achieve the fundamental objectives. The classification was done separately for Poland and Sweden.

The WITI (Why Is That Important) test [11] was used to classify the objectives. During this test, each objective was investigated by asking a question "Why is this objective important in the decision context?". If the answer is that a given objective is important because of its implications for some other objective, the objective was classified as a means objective; otherwise, it was classified as a fundamental objective.

4 Results

In the following sections, we present value-based objectives essential for the implementation of ICT for active and healthy ageing in Sweden and Poland. Based on the interviews conducted with Polish seniors, six fundamental and nine means objectives were identified as important for implementation of ICT for active and healthy ageing in the Polish context. In the Swedish context, we found six fundamental and twelve means objectives.

4.1 Fundamental and Means Objectives in Poland

In the following, there are objectives elicited on the basis of the interviews conducted with Polish respondents. The fundamental objectives include the following issues:

- *F1: Ensure acceptance by seniors* – The respondents emphasized that older adults should reveal positive attitudes towards technology and have some exposure to technology-based solutions. It is important to have an opportunity to test the solution and to learn how it operates. An accessible instruction manual should be available. The elderly should give their consent to use the solution. The elderly should not be afraid of technology and should have confidence in the solution.
- *F2: Increase seniors' activity* – The solution should support seniors' activity and help them to have more spare time. The solution should also help the seniors to thrive/develop themselves. The seniors should be involved in social life.
- *F3: Maximize mental comfort of seniors* – An older adult should have a sense of security, should not feel anxiety nor be lost. An older adult should feel needed and be appreciated. An older adult should not be afraid of being alone at home. The elderly should not be treated as a burden.
- *F4: Increase independence of seniors* – The solution should foster self-dependence of the elderly and should help them not to be a burden for the family. The elderly should be aware of the condition of their health and should be able to check their health condition unaided. The solution should help the elderly to improve their

health condition. The elderly should be able to operate the solution unaided. The solution should help the elderly not to get lost. The seniors should be able to live independently in their homes.

- *F5: Maximize usefulness for family and carers* – The solution should help to ensure security of other people. The solution should be helpful for the family and carers. There should not be a necessity to visit an older adult every day. The family should be relieved of some duties and have more spare time. The family should be confident of an older adult. The family should not have feelings of guilt while using the solution.
- *F6: Ensure support for the economy* – The solution should help national economy to develop. Companies delivering solutions supporting an active and healthy ageing should be established. The society should develop with the help of technology.

There are several means objectives elicited from the respondent answers and listed in the following.

- *M1: Ensure alignment with seniors' needs* – The solution should be easy to operate, comfortable, and adjusted to the older people's needs. The solution should be helpful for the elderly. There should be a possibility to use the solution at home. The solution should be able to react without human operation, and an older adult should not be forced to remember to bring it with him/her. The solution should be helpful in daily life. M1 supports F1, F5 and M2.
- *M2: Ensure personal contact with people* – The solution should help the elderly to have personal contact with family and other people. The elderly should have support in family; they should not be lonely. The seniors should be able to communicate with other people and should not be excluded from the society. M2 supports F2 and F3.
- *M3: Ensure privacy and data security* – The solution should guarantee data security and should ensure people's privacy. M3 supports F1, F3, F5 and M8.
- *M4: Maximize solution reliability* – The solution should be reliable and should operate quickly. The solution should undergo regular inspections. M4 supports F1, F3, F5, M5 and M8.
- *M5: Improve the quality of care* – The solution should provide fast information on the condition of health. It is essential to receive a fast and timely help. The solution should ensure better prevention and 24/7 care. Care should be easier, more accessible, less time consuming, and cheap. The solution should help to minimize health appointments. M5 supports F5, M6 and M8.
- *M6: Ensure access to various kinds of care* – It is important not to be completely dependent on the solution but also have access to traditional (involving people) type of care. The elderly should have access to free care. Technology solution should not supersede individual carers. The elderly should have access to professionals. Access to the alternative solution should be guaranteed. M6 supports F3, F4, M2 and M5.
- *M7: Maximize solution availability* – The solution should be widely available, cheap, and financially accessible for anyone. The solution should be subsidized by the government. It is essential to popularize the solution and have Internet access at home. M7 supports F1, F5, M6 and M8.

- *M8: Ensure usefulness for medical services* – The solution should be helpful for medical services and help them to operate efficiently. M8 supports F6, M5 and M6.
- *M9: Raise awareness of technology impact* – The solution should not cause negative social consequences, such as insensitivity. People should be ready to help the elderly and should understand potential consequences of loneliness. There is a need to make people aware of the consequences of the solution misuse. People's positive attitudes should be promoted. M3 supports F1, F3, F5, F6, M2, M3 and M6.

4.2 Fundamental and Means Objectives in Sweden

The following text summarizes the fundamental and means objectives elicited from interviews with Swedish respondents. The identified fundamental objectives are:

- *F1: Enhance seniors' digital inclusion*. Generally, the seniors revealed a positive attitude to digitalization, but they stressed that if special precautions are not taken, they can easily end up outside the digital community. Thus it is important to offer an opportunity for the seniors to keep up with technology by providing information, increasing awareness, develop ICT skills and involving seniors in development of ICT solutions.
- *F2: Maximize autonomy*. The seniors emphasized the importance of being able to decide about all aspects of their lives, i.e. where they live (at home, at an institution), what kind of help they get (digital or traditional), and to what extent they reveal information about themselves (balance between privacy and safety). For this reason, it is important to empower seniors in relation to other stakeholders (i.e., developers, companies, healthcare professionals) and increase their knowledge about existing solutions and consequences of their implementation so that they can make informed decisions in this context.
- *F3: Maximize mental well-being*. Well-being means that people feel safe, and needed. It also means that they do not feel like a burden for the society and their relatives. The seniors stressed that the new ICT solutions could positively impact these aspects and subsequently improve their well-being.
- *F4: Ensure acceptance*. To ensure acceptance of ICT solutions by the seniors, it is important that the solutions are useful (addressing the senior's actual needs) and that various solutions are available (seniors can choose among various solutions). The solutions should not be obtrusive. Further, the seniors emphasized the importance of sufficient support and education, so they are not afraid to use the new ICT solutions. It is also important that they reveal self-confidence and control as users of ICT. Finally, they emphasized the importance of trust in the new technology.
- *F5: Ensure equality and justice*. Swedish seniors stressed that everyone should have the same right and possibility to use and access the new solutions. They argued further that everyone should get the same information and education to be able to choose what is right for him/her. It is also important that the new solutions are equally affordable to all seniors.
- *F6: Enhance seniors' activity*. The seniors underlined the important role of ICT in enhancing their activity. They stressed, for instance, that apps and/or social robots

could inspire them to physical activity. They also pointed out that ICT solutions could enhance their involvement in different interest groups, interaction with other people, and political discussion. The seniors stressed further that it is fun to learn new things related to ICT (self-development) and that those who are skilled could help others.

The means objectives elicited from the respondent answers are listed in the following.

- *M1: Create awareness of ICT's impact on society.* Careless implementation of ICT can have a negative impact on seniors' lives, dignity, and well-being. Therefore long-term social consequences of ICT implementations should be considered and analyzed. The seniors pointed out that ICT may change our way of thinking, professional roles, family relationships, etc. Thus the new ICT solutions should be implemented thoughtfully, based on informed decisions. M1 supports all fundamental objectives.
- *M2: Provide possibility for ageing in place.* Our participants emphasized the importance of being able to age at home in well-known settings, close to their friends and neighbors, among things and places they like (as opposed to moving to a care institution). They believed that implementation of new ICT solutions could make it possible. M2 supports F3 as well as M5.
- *M3: Enhance safety.* The new ICT solutions have also been seen as means to enhance seniors' safety. Our respondents stated that it is important that the solutions can detect fall, raise an alarm and facilitate communication with family and cares. M3 supports F3, M2, and M7.
- *M4: Maximize solution's availability.* Our respondents pointed out that there is a need for developing relatively cheap and simple everyday solutions that seniors could afford. For the other, more complex and expensive solutions (such as social robots, health care systems, etc.) sufficient payment models need to be developed. Generally, such solutions should be provided by municipalities together with sufficient training and support. M4 supports F3–F6.
- *M5: Minimize loneliness.* Implementation of the new ICT solutions is seen as a means to reduce loneliness by enhancing social contact and supporting social interaction. In this context the seniors mentioned online games designed for their needs, interactive social robots, and different kind of communications tools that would help them to be closer to their family and friends. However, the seniors also emphasized the importance of physical contact with other people, "to have someone to share daily activities with." M5 supports F3, M2, and M7.
- *M6: Ensure alignment with seniors needs.* It is important that the new ICT solutions are designed with the seniors' needs and requirements in mind. The solutions need to be usable (ease of use, easy to learn, simple, with a friendly design). They should be flexible (possible to adjust). Many different solutions should exist so it is possible to choose the most suitable for each situation. Still, traditional ways of doing things should be possible to choose if so required. M6 supports F4, F5, M2 and M7.
- *M7: Increase independence of seniors.* It is important to develop solutions that help seniors to perform daily activities at home; especially if the inhabitant suffers from

disabilities related to ageing or health condition. The respondents also stressed that it is important to be and feel healthy. M7 supports F3, F7.

- *M8: Increase knowledge and awareness.* It is important that seniors' are informed about existing solutions. Their awareness and knowledge within this area can be increased by workshops, folders and by including them in development and trials of ICT solutions. The seniors also emphasized their own role in increasing awareness about the new solutions among friends and relatives. M8 supports F1–F4.
- *M9: Ensure privacy and data security.* It is important to consider both the seniors' and their partners' privacy when the new ICT solutions are implemented in their homes. Seniors want to be informed about the extent of monitoring, as well as how the information is handled, and who can access the information. It is also important to ensure correct and careful interpretation of the collected data. M9 supports F2, F3, and F4.
- *M10: Maximize reliability of the solution.* It is important that the solutions work without any technical errors and show correct data. Alternative ways of doing things must exist in case of emergency such as power cut. Also someone must respond to alarms otherwise the solutions is not reliable even if it works without technical errors. M10 supports F4, F6, M2 and M3.
- *M11: Promote digital development.* It is important that municipalities invest in development of ICT in various areas and that the solutions are implemented in future homes and care institutions as standard equipment. M11 supports F5, M2, M4, M7 and M12.
- *M12: Improve the quality of care.* The seniors pointed out the importance of using ICT for more efficient use of existing resources (personnel's time), for improving accessibility to care professionals and facilitating face-to-face interaction. M12 supports M2, M3, M5 and M7.

5 Discussion

The findings of the current study, obtained as a result of a VFT-based analysis, revealed both similarities and differences in fundamental and means objectives between the two investigated countries. The health and elderly care system in Poland requires a significant involvement of family or caregivers hired by the family in terms of financial and time-related resources. Therefore, Polish seniors expect that ICT solutions would support the family in care provision. However, the seniors' expectations for new technology also raise fears of losing contact with their relatives. Therefore, it is important for Polish seniors that new technologies ease access to various kinds of care, in particular those financed by the government, without eliminating traditional care.

In Sweden, access to elderly care is guaranteed for every citizen and, additionally, the level of technological development is relatively high. Therefore, the Swedish seniors' expectations focus on the issues related to equality and justice in access to modern solutions. The expectations of Swedes concern the need for even greater enhancement of seniors' digital inclusion. They emphasize the importance of the solution's alignment with individual needs of seniors, their involvement in the design process, and the possibility of using the solutions in various domains of daily life, including entertainment.

The last aspect is associated with differences in seniors' health status between Poland and Sweden. Seniors at the same age in Sweden are healthier than Polish seniors; the latter also need access to healthcare earlier in their lives. For Poles, new solutions should be helpful mainly in health-related matters (i.e., access to care, health maintenance, and prevention); therefore, solution reliability is the most crucial issue. For Swedes, using technology to participate in social life and maintain contact with friends appears the equally important issue. In their opinion, it is essential to minimize loneliness by developing enjoyable solutions (games, interactive robots) or facilitating different kinds of communication (video, voice).

The last fundamental objectives differently perceived by Polish and Swedish respondents refer to seniors' independence and autonomy. In particular, Polish seniors expect that new technology-based solutions would help them to be independent. They perceive independence mainly as being self-reliant. They expect technology to facilitate their everyday, independent life, so that they do not have to ask others for help and are able to monitor their health status on their own. For Swedes, in turn, autonomy is a very important issue. They perceive autonomy as their right and part of their dignity. Therefore, they expect to achieve greater empowerment in relation to other stakeholders, e.g., by being informed about the level of privacy and data security.

Some objectives are very similar for seniors in Poland and Sweden. The most important such objectives include ensuring acceptance and alignment with seniors needs. The common features of solutions for seniors in Poland and Sweden are their usability and usefulness. It should also be possible to adjust the solutions to individual needs addressing the senior's actual situation (health condition, IT competencies, place of residence). Further, the seniors emphasized the importance of education to ensure that they are not afraid of using these new ICT solutions and that they have confidence in them.

Other important common goals include raising awareness of technology impact on society, focusing on problems related to ageing and the role of ICT. The seniors pointed out that ICT may change our way of thinking and family relationships. It is therefore important to build a sense of security and awareness that using ICT for elderly care is just a supporting way of caring, not a replacement for individual care.

This research has several significant implications for analysis and design of ICT solutions for active and health ageing in Poland and Sweden.

First, it is important to involve seniors in the development and implementation of ICT solutions for active and healthy ageing. Our study shows that seniors are both willing and capable of expressing significant requirements in relation to modern ICT solutions. As shown in the result section, acceptance of the new solutions by senior-users requires understanding of their individual needs and actual situation related to their health condition, IT competencies, and place of residence. Both Polish and Swedish seniors also emphasized the importance of an opportunity to test the solution and to learn how it operates before actually using it. Therefore, we believe that they should be involved in the development and implementation of ICT solutions for active and healthy ageing.

Second, the development of ICT solutions for active and healthy ageing should apply a broad perspective considering aspects beyond technical features and functionality. Our study shows that technology acceptance by seniors not only depends on the technical features and functionality, but equally important is establishing of supporting structures, awareness programs and training. Only then it is possible to create trust, acceptance and positive attitude towards technology among senior-users.

Third, in order to be able to successfully implement ICT solutions for active and healthy ageing, one must understand and consider the social implications the technology implies for the various stakeholders and the society. Many of the existing projects aiming to provide technology supporting older people have a narrow technical focus ignoring the complex social context in which the technologies are implemented and used [14]. Our interviewees highlighted several challenges in this context and argued that modern ICT solutions should be implemented thoughtfully and based on informed decisions to avoid irreversible, negative social consequences concerning the individual user, other stakeholder groups and the society as a whole.

Fourth, the socio-economic differences between different countries should be acknowledged when ICT solutions for active and healthy ageing are developed and implemented. Our study shows that Polish and Swedish seniors emphasize different objectives in this context. This means that strategies and ICT solutions for active and healthy ageing cannot be just transferred between countries in Europe without reflection and adaptation of the solutions to the specific country context.

6 Conclusion

By applying Value-focused thinking [11] this study investigated objectives important for the implementation of ICT for active and healthy ageing according to Swedish and Polish seniors. For Polish seniors it is important that new technology ensures better access to different kinds of care by faster access to fuller information and facilitating communication with the family and healthcare professionals, thus reducing the cost of care. Seniors in Sweden, in turn, expect equality and justice in access to new solutions, which should not only support health maintenance, but also facilitate involvement in social life and increase their digital inclusion. This research has several significant implications for analysis and design of ICT solutions for active and healthy ageing in Poland and Sweden: (1) First, it is important to involve seniors in the development and implementation of ICT solutions in this context, (2) the development of ICT solutions for active and healthy ageing should apply a broad, holistic perspective considering aspects beyond technical features and functionality, (3) in order to be able to successfully implement ICT solutions for active and healthy ageing, one must understand and consider the social implications the technology implies for the various stakeholders and the society, (4) the socio-economic differences between different countries in Europe should be acknowledged when ICT solutions for active and healthy ageing are being developed and implemented. In the end it should be emphasized that this research is an exploratory study based on a limited number of interviews. Therefore, further research is required in order to supplement the sample and perform a more in-depth data analysis.

Acknowledgments. This research has been financed in part by the funds granted to the Faculty of Management, Cracow University of Economics, Poland, within the subsidy for maintaining research potential.

References

1. Bloom, D.E., et al.: Macroeconomic implications of population ageing and selected policy responses. Lancet **385**(9968), 649–657 (2015)
2. European Commission: Access to healthcare and long-term care: equal for women and men? Final synthesis report. Publications Office of the European Union, Luxembourg (2010)
3. European Commission: The 2015 ageing report. Economic and budgetary projections for the 28 EU member states (2013–2060). European Economy 3 (2015)
4. European Commission: The 2018 ageing report. Underlying assumptions and projection methodologies. European Economy Institutional Paper 065 (2017)
5. Eurostat (2017). http://ec.europa.eu/eurostat/data/database
6. Foster, L., Walker, A.: Active and successful aging: a European policy perspective. Gerontologist **55**(1), 83–90 (2014)
7. Gregory, R., Keeney, R.L.: Creating policy alternatives using stakeholder values. Manag. Sci. **40**(8), 1035–1048 (1994)
8. Harper, S.: Economic and social implications of aging societies. Science **346**(6209), 587–591 (2014)
9. Iakovidis, I.: The European dimension of eHealth: challenges of innovation. Int. J. Healthc. Manag. **5**(4), 223–230 (2012)
10. Kapadia, V., Ariani, A., Li, J., Ray, P.K.: Emerging ICT implementation issues in aged care. Int. J. Med. Inform. **84**, 892–900 (2015)
11. Keeney, R.L.: Value-Focused Thinking. Harvard University Press, Cambridge (1992)
12. Keeney, R.L.: Creativity in decision making with value-focused thinking. Sloan Manag. Rev. **35**, 33–41 (1994)
13. Klimczuk, A.: Comparative analysis of national and regional models of the silver economy in the European Union. Int. J. Ageing Later Life **10**(2), 31–59 (2016)
14. Kolkowska, E., Avatare Nöu, A., Sjölinder, M., Scandurra, I.: Socio-technical challenges in implementation of monitoring technologies in elderly care. In: Zhou, J., Salvendy, G. (eds.) ITAP 2016. LNCS, vol. 9755, pp. 45–56. Springer, Cham (2016). https://doi.org/10.1007/978-3-319-39949-2_5
15. Kolkowska, E., Soja, E.: Attitudes towards ICT solutions for independent living among older adults in Sweden and Poland: a preliminary study. In: Kowal, J., Kuzio, A., Mäkiö, J., Paliwoda-Pękosz, G., Soja, P., Sonntag, R. (eds.) 2017 Proceedings of the International Conference on ICT Management for Global Competitiveness and Economic Growth in Emerging Economies, ICTM 5/2017, Wrocław, Poland, 23–24 October 2017, pp. 35–45. University of Wrocław, Wrocław (2017)
16. Peek, S.T.M., et al.: Older adults' reasons for using technology while aging in place. Gerontology **62**(2), 226–237 (2016)
17. Potocnik, K.: Healthy ageing and well-being at work. In: Parry, E., McCarthy, J. (eds.) The Palgrave Handbook of Age Diversity and Work, pp. 171–194. Palgrave Macmillan, London (2017). https://doi.org/10.1057/978-1-137-46781-2_8
18. Social Protection Committee and European Commission: Adequate social protection for long-term care needs in an ageing society. Publications Office of the European Union, Luxembourg (2014)

19. Soja, E.: Information and communication technology in active and healthy ageing: exploring risks from multi-generation perspective. Inf. Syst. Manag. **34**(4), 320–332 (2017)
20. Soja, E.: Supporting active ageing: challenges and opportunities for information and communication technology. Zarządzanie i Finanse J. Manag. Finan. **15**(1/2017), 109–125 (2017)
21. Walker, A., Maltby, T.: Active ageing: a strategic policy solution to demographic ageing in the European Union. Int. J. Soc. Welf. **21**(1), 117–130 (2012)
22. World Health Organization (WHO): World report on ageing and health. World Health Organization, Geneva (2015)
23. Zacher, H., Kooij, D.T.A.M., Beier, M.E.: Successful aging at work: empirical and methodological advancements. Work Aging Retire. **4**(2), 123–128 (2018)

Analysing Opportunities and Challenges of Integrated Blockchain Technologies in Healthcare

Ebru Gökalp$^{(\boxtimes)}$, Mert Onuralp Gökalp, Selin Çoban,
and P. Erhan Eren

Informatics Institute, Middle East Technical University, Ankara, Turkey
{egokalp,gmert,coban.selin,ereren}@metu.edu.tr

Abstract. Blockchain is a disruptive technology with the potential to have a significant impact on business models and industries, similar to the adoption of Internet. Blockchain promotes distributed, open, inclusive, immutable, and secure architectural approaches, instead of centralized, hidden, exclusive, and alterable alternatives. The adoption of blockchain in the healthcare domain offers promising solutions for securing communications among stakeholders, efficient delivery of clinical reports, and integrating various kinds of private health records of individuals on a secure infrastructure. Accordingly, the main aim of this study is to propose a holistic blockchain structure covering all stakeholders in the healthcare domain and to analyse opportunities and challenges by presenting an integrated blockchain architecture. The comprehensive view of blockchain based healthcare system consists of services as follows: personal medical health record storage and access, personal genomic data storage and access, inventory tracking and buy-sell mechanism, health research commons, health document notary services, doctor services, digital health wallet, peer-to peer insurance. The opportunities of using blockchain in the healthcare domain are considered with respect to several viewpoints such as transparency, accountability, decentralization, record accuracy, secure transactions, interoperability, lower costs, collaboration, agility, individualized care with specialized treatment, improved diagnosis methods, risk of insurance contract, prevention of counterfeit drugs and improved quality of medical research. Challenges associated with the implementation of blockchain in the healthcare domain are also highlighted, such as governance, lack of legacy, privacy, sustainability, scalability, adoption of participation, and cost of operations.

Keywords: Blockchain · Healthcare · Holistic approach · System design

1 Introduction

Blockchain is a sort of dispersed ledger of cryptographically chained blocks where value exchange transactions are consecutively aggregated. Every block is immutably recorded across peer-to-peer (P2P) networks, and each of them is chained to the former block using cryptographic assurance systems. This transaction ledger is built in a

© Springer Nature Switzerland AG 2018
S. Wrycza and J. Maślankowski (Eds.): SIGSAND/PLAIS 2018, LNBIP 333, pp. 174–183, 2018.
https://doi.org/10.1007/978-3-030-00060-8_13

decentralized P2P network structure, so that any kind of transactions and assets does not need to deal with a centralized intermediary. With the help of emerging blockchain technology, any of the digital assets can be transacted, encoded, verified, and maintained in a more efficient way compared to the present situation including personal health records, financial transactions, and confidential documents.

The key innovation of blockchain technology is its architecture comprised of decentralized secure transaction systems. It allows decentralization and disintermediation of all transactions of any type between all stakeholders on a common universal basis. Blockchain behaves like an application layer to progress internet protocols stack, appending a new tier to the internet to allow economic transactions, and immediate digital currency payments [1].

We all witness the impact and growth of information and its value. The convergence of advanced computing and communications technologies triggers the first stage of the network of information, the second stage will be leveraged by cryptography, mathematics, software engineering, behavioural economics, and their smart assembly. In this context, blockchain has the potential to turn business models upside-down and transform industries as Internet did. It appears to challenge the structure of society where we have defined values and rewarded participation. Blockchain promotes a distributed, open, inclusive, immutable and secure architecture in exchange for centralized, hidden, exclusive and alterable architecture. To this end, blockchain enables us to create and trade value in societies. Through blockchain technology, confidential and verified transactions can be carried out directly among third-parties by mass collaboration. Rather than these, in today's world, corporations are motivated by profit or governments are motivated by power. Thus, a true paradigm which is the peak of what Alan Turing [2] stated earlier, may now be feasible with this decentralized ledger technology.

Considering blockchain's feasibility, service related areas such as healthcare sector are beginning to change and adapt themselves to integrate their current status with the blockchain technology. The business value-add is expected to increase by more than $176 billion by 2025, and exceed $3.1 trillion by 2030 [3]. According to the World Economic Forum report, 10% of global gross domestic product will be stored using the blockchain technology [4]. The global blockchain market compound annual growth rate (CAGR) is anticipated to rise 71.46% between 2017 and 2022 to reach a total market size of US$4.401 billion by 2022, increasing from US$0.297 billion in 2017 [5]. These estimates do not astonish investors who invest in blockchain start-ups, but the potential of blockchain is still unknown throughout society. Nevertheless, this technology is already being adopted by industries such as banking, minerals and mining, and food procurement [5, 6]. It is now discussed to be utilized in the healthcare domain to revolutionize health information technology [7] and its payment models [8, 9].

In the healthcare domain, researchers and professionals compete against fragmented data, poor communications and medical workflows with missing parts because of vendor specific and incompatible health systems. Moreover, health IT systems are influenced by the chaos of sharing medical and financial data, thus they suffer from critical faults related to privacy and security. All of these make individualized patient care difficult [7].

The main advantage of blockchain is that it provides solutions for disclosure and accountability issues between corporations and individuals in terms of both side's self-interests. The adoption of blockchain technology in the health domain offers promising solutions for securing communications among stakeholders, efficient delivery of clinical reports, and integrating various kinds of private health records of individuals on a secure infrastructure.

Accordingly, the main focus of this study is to propose a holistic blockchain structure covering all stakeholders in the healthcare domain and to analyse the opportunities and challenges by revealing integrated blockchain technologies.

The rest of the study is organized as follows; the related works are reviewed in Sect. 2. In Sect. 3, a blockchain framework for the healthcare domain is proposed and this framework is explained with respect to its usage areas. Opportunities and challenges of blockchain technology and the proposed framework are discussed in Sect. 4. Then we conclude our study and give future directions in Sect. 5.

2   Related Works

The significant potential of blockchain technology has sparked a growing research interest in various domains including healthcare. It offers a large field of application for the health domain as a result of promised decentralized management, immutable audit trail, data provenance, security, privacy and robustness [10]. However, since it is an emerging technology, there is a limited number of studies in the literature that unifies blockchain technology with health domain.

The recent studies [11–13] discuss improving medical record management by handling secure data transactions, exchanging private health data among stakeholders and its adoption in existing infrastructures. The study of Yue et al. [11] proposes a blockchain architecture which enables patients to control and share their own private health related data easily and securely with third parties. Similarly, the study of Peterson et al. [12] offers a blockchain based approach to share health related data securely. Health-Bank [8] also offers an approach to store and manage data from mobile health applications and wearables.

Another important contribution of blockchain technologies in the health domain is improving and accelerating health related research by standardizing databases of health information and providing non-identifiable patient information as age, gender, illness, accessible to researchers [1, 14]. The recent study of Linn and Koo [15] describes a blockchain based database access control management system to promote the development of medicines and advance medical research. Another study, MedRec [16], provides a prototype implementation to demonstrate how blockchain architecture can contribute to the improvement of health data quality and quantity for medical research. These studies also state that patients are more prone to share their personal health data with a secure blockchain system.

Blockchain also helps to verify the financial transactions for health insurance and confirm health information documents which include test results, prescriptions, and health reports [17]. The Health-Coin is offered for health-related financial transactions and its contribution is showed for the case of a cure of Type 2 diabetes in [9]. Another study [18] proposes the use of blockchain technology in combination with digital signatures to create smart digital health contracts.

To sum up, there are a limited number of studies in the literature that combines blockchain technology with health domain. These studies primarily focus on proposing a solution for specific cases. Therefore, they mainly lack providing a comprehensive investigation of blockchain technologies to be applied in health domain. To this end, in this study, we examine the relationships between blockchain and health domain from sociotechnical perspective and propose a holistic approach that integrates applications of blockchain technologies in health domain as well as analyse the opportunities and challenges to provide a comprehensive picture.

3 Blockchain Technology in Healthcare

The proposed comprehensive framework of blockchain based healthcare system is given in Fig. 1. In the proposed framework, patient health records including patient and clinic generated data, drug records as well as medical device records are digital assets that are stored and accessed via blockchain infrastructure. These digital assets are encrypted and digitally signed as data lake to provide secure personal medical health record storage and access. Seamless exchange of data and use of the exchanged data across health organizations and application vendors occur with a secure underlying infrastructure. Consistent complete history of timestamped records can be used by the applications including smart contracts, health analytics, and network of medical technologies for users of patients, caregivers, medical device manufacturers, researchers, insurance companies, hospital, pharma companies as well as regulatory companies.

The services provided by the proposed framework are personal medical health record storage and access, personal genomic data, supply chain management of hospital assets, smart contract technology, private blockchain technology, health research commons, health document notary services, and doctor services.

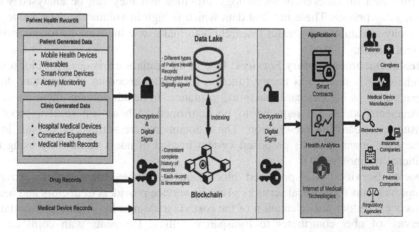

Fig. 1. Comprehensive framework of blockchain based healthcare system

- **Personal Medical Health Record Storage and Access:** Medical Health Records (MHR) provide seamless communication and exchange of data, among organizations and application vendors in health sector.
- **Personal Genomic Data:** The blockchain technology helps to aggregate DNA data from different holders and sources without the need to collect it into a central database and to back-up DNA of the person to spread personal genetic material to many computers around the world. Moreover, the utilization of blockchain architecture enables us to tag, track and cross-reference private health related data with a secure communication infrastructure as well as to provide access to large sets of genetic data with correlated clinical records, while keeping anonymity and privacy, as dna.bits aims to achieve [1].
- **Supply Chain Management of Hospital Assets:** The blockchain technology can also be leveraged in supply chain management systems to manage inventory of hospitals and to regulate buy-sell mechanism for all hard-medical assets.
- **Smart Contract Technology:** Smart contract technology has the potential of reducing insurance fraud at every level from corporate management of funds down to individual claims and it can do all of this with less costs due to less overhead required to maintain and operate the system. It mainly serves as a legal digital agent to create, store and enforce contracts by supporting data exchange in a real-time manner with the blockchain architecture.
- **Private Blockchain Technology:** Private blockchain requires permission to read the information on the blockchain. That sets the parties who can transact on the blockchain and write new blocks into the chain. Patients give permissions to doctors and other parties to access records. Various access levels can be described to control who can access records, who can modify them, and who has the ultimate authority in the system based on business agreements and bureaucracy. The combination of security and flexibility is more suitable in healthcare than transparent public blockchain.
- **Health Research Commons:** Aggregating patients' medical records, genome and connectome files and quantified self-data commons in an encrypted pseudonymous form based on blockchain technology provides that they can be analysed while remaining private. These medical data which is high in volume, variety, value and veracity is valuable for health researchers to make new innovations for improving public health.
- **Health Document Notary Services:** One of the features envisioned for blockchain technology is that it has notary function for digital encoding of important documents such as identity card, passport, insurance. In health domain, there are some documents which are required to be confirmed such as test results, proof of insurance, treatment, prescription. These documents are verified in seconds in an encrypted manner in the proposed system instead of hours or days by using traditional technology [1].
- **Doctor Services:** The proposed blockchain based healthcare system provides improvements for medical services given by service providers of doctors and health practices. Automated calculation of the cost via trade-nets like online transportation service of uber contributes to transparent billing. Payments with coins on the blockchain technology provide efficient payment process and interactivity.

4   Discussion

In this section, we review the opportunities and challenges of utilizing blockchain technology in healthcare domain.

4.1   Opportunities of Blockchain Technology in Healthcare

The opportunities of using blockchain in healthcare domain are considered with respect to several viewpoints such as transparency, accountability, decentralization, record accuracy, secure transactions, interoperability, lowering costs, collaboration and agility. Integration of health data with blockchain technology is illustrated in Fig. 2.

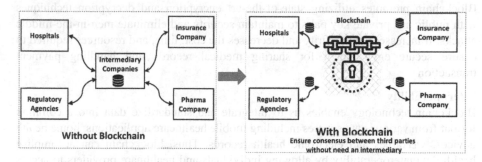

Fig. 2. Integration of health data with blockchain technology

Transparency

The decentralization feature of blockchain allows us to create immutable and objective data resources where individuals can retrieve up-to-date information, statistics, and reviews about healthcare service providers. Moreover, with the help of patient-centered care [19] and secure peer to peer network, patients can freely share their private health records with medical providers without the worry of the information being stolen. Thus, individuals are empowered to make an effective choice in their healthcare. Empowered patients are more likely to have less anxiety regarding healthcare provider interactions, and to make better, more informed choices about their healthcare. Moreover, the entire treatment of a patient is handled transparently, as provided by the flow of information among stakeholders resulting in improved confidence.

Accountability

The insurance and reimbursement process is one of the most infuriating and complex aspects of the healthcare industry. Insurance companies and governments do not want to be overbilled or fraudulently billed by doctors seeking to provide medically unnecessary procedures. The blockchain technology can make documentation and sharing much easier. Thus, all parties involved in healthcare need to be more accountable as the medical information becomes more transparent, shareable and verifiable.

Record Accuracy

In current health information systems, patients may encounter inaccuracies in their health records such as their health history may not be shared with all stakeholders. Therefore, the data consistency among healthcare service providers is an important challenge with the traditional health information management technologies. With the help of blockchain, patients' health records can be stored in disparate systems and every member of the chain is connected and data is updated in all of the computers at once. This improved communication and record accuracy could significantly reduce common issues like medication errors. Patients are able to have greater trust in their providers when accurate and up-to-date information is available for all members of a healthcare blockchain.

Secure Transactions

Blockchain promotes utilizing state-of-the-art encryption and decryption techniques with public and private key pairs to maintain security and eliminate men-in-the-middle attacks. It increases data security and decreases the cost, time, and resources required to ensure secure environment for sharing medical records and handling payment transaction.

Interoperability

Blockchain technology enables us to integrate and standardize data into a common format from various data sources including mobile healthcare applications, home health devices, wearables and electronic health records. Thus, blockchain can accomplish health data interoperability by allowing individuals and healthcare providers to access data across different systems such as DNA bank, organ donor-ship and blood banks. This interoperability also increases the speed to retrieve relevant information from different data sources and increases data quality and quantity for medical researches. The emerging big-data analytics, machine learning and deep learning technologies can be utilized on blockchain data to further analyse the intersection of demographics, genetic markers, and a range of other data sets for research and development purposes.

Lower Cost

Blockchain makes documented and accurate information available to patients, thereby enhancing their decision making capability with respect to the financial aspects of the treatments. The patients can keep track of their expenses and they become aware of their spending in relation to the services offered by the providers. The availability of blockchain also reduces the costs with respect to the operation and maintenance of the providers' transactional systems. Furthermore, the care plan of patients can be updated in a real-time fashion based on health records in the blockchain by using smart contracts. For instance, if a patient is frequently checked by Emergency Department (ER) in a short period of time, the insurance company may add a caveat in the care plan to investigate the patient health status to avoid frequent ER re-admittance. This type of smart contract aims to reduce burdens of the ER while reducing the overall healthcare costs of a patient.

Collaboration

Blockchain also provides collaboration between producers, suppliers, retailers, and end customers to gather historical events and actions throughout the chain such that individuals and service providers can access information transparently about health products, drugs, MRI devices, surgery materials and so on. The importance of collaboration rises because according to the World Health Organization, 10% of drugs including antibiotics, painkillers, contraceptives and other prescription drugs, are counterfeit worldwide, and the number rises up to 30% in developing countries [20]. When no trace of the forged drug and its expected activity is considered, it is extremely jeopardous for the human population [20].

Agility
With the availability of mobile health devices and wearables, it is possible to collect health related records from patients in a real-time fashion for 24 h and 7 days. Personalized diagnoses and treatments can be generated by turning these collected data into meaningful and valuable inputs. Additionally, emergency medical situations can be handled by monitoring real-time health records to detect unusual changes in vital parameters. Furthermore, it also allows detection and containing of epidemic disease earlier by rapidly discovering, isolating and driving change for environmental conditions that impact public health.

4.2 Challenges of the Application of Blockchain Technology in Healthcare

Governance
Interoperability across public and private blockchain is required for the proposed blockchain based healthcare system. This raises the need of standards and agreements in a globally coordinated manner across international borders and jurisdictions.

Lack of Legislation
There are difficulties in preparation of appropriate regulations for governance rights of ownership regarding medical transactions around the world for the proposed blockchain based healthcare system. Due to the existence of many stakeholders, ownership of data and existing medical law of traditional healthcare system are important issues that need to be addressed appropriately. Modifying current regulatory framework according to new administration policy objectives governing digitally described, automated and universal nature of blockchain will be difficult. Ownership of records, granted access rights, and distributed storage structure of blockchain should be clarified carefully.

Transparency and Privacy
Blockchain technology highlights transparency which may not be desirable in some cases for the health domain. Although it provides security through encryption, availability of a database, even in encrypted form, is seen as an important issue for healthcare stakeholders. Therefore, access control in the context of blockchain should be addressed properly.

Sustainability
Encryption key plays an important role in blockchain. There is no way to recover the private encrypted key. This brings complexity in healthcare data due to its long-lasting

nature. Missing parts in a patient's health record significantly reduce its reliability and value. Additionally, hacking or stealing a user's private encryption key enables access to all information stored in relation to the user.

Scalability

As users add data, blockchain grows—in this case, by storing all of the hashes associated with the appended data. This increases storage and computational power demands, which means the network might have fewer nodes with enough computing power to process and validate information on the blockchain. If health professionals fail to meet storage and computational power demands, the potential for increased centralization and slower data validation and confirmation grows.

Adoption for Participation

Blockchain technology requires a network of interconnected computers for supplying the necessary computing power for both creating blocks of a transaction and cryptocurrency. Participants should be encouraged through incentive mechanisms for contributing to computing power. Additionally, health organizations may need to be encouraged for adopting the blockchain technology and participating in the shared network. The impact of blockchain will increase with the number of participating entities.

Cost of Operation

While the cost of setting up and operating such a system as well as migration from traditional health information systems is not yet known, open-source technologies and the distributed nature of blockchain can help reduce it. Continuous operation and maintenance of the proposed blockchain based integrated health system requires constant availability of resources for troubleshooting, updating, backup, and reporting purposes.

5   Conclusion

Blockchain provides various advantages such as decentralized architecture together with transparent and immutable structure, which may prove to be beneficial for the health domain. While it presents significant opportunities, it brings some challenges in the form of governance issues, lack of legislation, trust and privacy, scalability, adoption for participation, and cost of operation. Accordingly, the opportunities and challenges of the application of blockchain technology in healthcare domain are investigated from a holistic viewpoint in this study.

The healthcare sector is characterized by the major size of investments and spending, inefficient operations, and problematic legislations around the world. Blockchain allows us to address some of these issues by regulating the communication among stakeholders and to drive new digital business models and health initiatives. To this end, blockchain has the potential to provide promising solutions in the near future and these solutions may result in disruptive changes in the healthcare industry. As part of future work, it is planned to expand this study by incorporating sample blockchain-based healthcare use cases.

References

1. Swan, M.: Blockchain: Blueprint for a New Economy. O'Reilly Media Inc., Newton (2015)
2. Alan, M.: Turing. Computing machinery and intelligence. Mind **59**(236), 433–460 (1950)
3. Gartner: Forecast: Blockchain Business Value, Worldwide, 2017–2030 (2017)
4. World Economic Forum: Deep Shift. Technology Tipping Points and Societal Impact (2015)
5. Research and Markets: Blockchain Market - Forecasts from 2017 to 2022 (2017)
6. ONC: Connecting health and care for the nation: a 10-year vision to achieve an interoperable health it infrastructure (2014)
7. Middleton, B., et al.: Enhancing patient safety and quality of care by improving the usability of electronic health record systems: recommendations from AMIA. J. Am. Med. Inform. Assoc. **20**(e1), e2–e8 (2013)
8. Mettler, M.: Blockchain technology in healthcare: the revolution starts here. In: 2016 IEEE 18th International Conference on e-Health Networking, Applications and Services, Healthcom 2016, pp. 1–3 (2016)
9. Basu, A., Subedi, P., Kamal-Bahl, S.: Financing a cure for diabetes in a multipayer environment. Value Health **19**(6), 861–868 (2016)
10. Kuo, T.T., Kim, H.E., Ohno-Machado, L.: Blockchain distributed ledger technologies for biomedical and health care applications. J. Am. Med. Inf. Assoc. **24**(6), 1211–1220 (2017)
11. Yue, X., Wang, H., Jin, D., Li, M., Jiang, W.: Healthcare data gateways: found healthcare intelligence on blockchain with novel privacy risk control. J. Med. Syst. **40**(10), 218 (2016)
12. Peterson, K., Deeduvanu, R., Kanjamala, P., Boles, K.: A blockchain-based approach to health information exchange networks (2016)
13. Xia, Q., Sifah, E.B., Asamoah, K.O., Gao, J., Du, X., Guizani, M.: MeDShare: trust-less medical data sharing among cloud service providers via blockchain. IEEE Access **5**, 14757–14767 (2017)
14. Benchoufi, M., Ravaud, P.: Blockchain technology for improving clinical research quality. Trials **18**(1), 335 (2017)
15. Linn, L., Koo, M.: Blockchain for health data and its potential use in health it and health care related research. In: ONC/NIST Use of Blockchain for Healthcare and Research Workshop. ONC/NIST, Gaithersburg (2016)
16. Ekblaw, A., Azaria, A., Halamka, J.D., Lippman, A.: A case study for blockchain in healthcare: 'MedRec' prototype for electronic health records and medical research data. In: Proceedings of IEEE Open & Big Data Conference (2016
17. Broader Perspective: Blockchain Health - Remunerative Health Data Commons & HealthCoin RFPs. http://futurememes.blogspot.com.tr/2014/09/blockchain-health-remunerative-health.html. Accessed 03 Dec 2017
18. Mytis-Gkometh, P., Drosatos, G., Efraimidis, P.S., Kaldoudi, E.: Notarization of knowledge retrieval from biomedical repositories using blockchain technology. In: Maglaveras, N., Chouvarda, I., de Carvalho, P. (eds.) Precision Medicine Powered by pHealth and Connected Health. IP, vol. 66, pp. 69–73. Springer, Singapore (2018). https://doi.org/10.1007/978-981-10-7419-6_12
19. Oates, J., Weston, W.W., Jordan, J.: The impact of patient-centered care on outcomes. Fam. Pract. **49**, 796–804 (2000)
20. Organization, W.H.: Growing threat from counterfeit medicines. Bull. World Health Organ. **88**(4), 247–248 (2010)

A Productivity Dashboard for Hospitals: An Empirical Study

Miguel Pestana[1], Rúben Pereira[2(✉)], and Sérgio Moro[2]

[1] Instituto Universitário de Lisboa (ISCTE-IUL), Lisbon, Portugal
mrcmp@icste-iul.pt
[2] ISTAR-IUL, Instituto Universitário de Lisboa (ISCTE-IUL), Lisbon, Portugal
{ruben.filipe.pereira, sergio.moro}@iscte-iul.pt

Abstract. Health information systems are key assets in managing health units' daily operations. Nevertheless, literature is scarce concerning information systems for increasing and managing hospital productivity. This study aims at filling such gap through an empirical research based on large Portuguese hospital.

Specifically, a dashboard prototype is proposed addressing productivity indicators in areas such as assistance, hospitalization, surgery, among others.

This dashboard is tuned using a design science research approach where health experts successively validate the prototype.

Interviews are conducted to assess the benefits of using our proposal to manage productivity on a daily basis.

Keywords: Health information systems · Business Intelligence
Dashboards · Hospital management · KPIs

1 Introduction

Healthcare is a sector with exponential growth, making it one of the largest industries in the world with enormous impact on countries' economy [1]. However, like all other institutions, healthcare organizations are at the mercy of the unstable environment of external factors such as technological advances. For these reasons and due to this impact on the healthcare system at the level of effectiveness it is necessary that hospitals constantly improve their performance [2]. Therefore, to improve performance, it is necessary to measure and evaluate it, so that it is possible to allow healthcare organizations to define strategies to achieve hospitals' goals [1].

The strategic objectives of healthcare organizations are difficult to achieve given their complexity [3]. The data volume is increasing due to more than 20 years of electronic storage of patients' data, with the Hospital Information System (HIS) being able to store and subsequently provide useful information throughout the hospital's medical history [4]. The impact of the introduction of information and communications technology (ICT) in the multifaceted health sector is well known and acknowledged by health stakeholders who have witnessed the role of ICTs in diverse levels in hospital organization [5].

© Springer Nature Switzerland AG 2018
S. Wrycza and J. Maślankowski (Eds.): SIGSAND/PLAIS 2018, LNBIP 333, pp. 184–199, 2018.
https://doi.org/10.1007/978-3-030-00060-8_14

Business Intelligence (BI) is a valuable asset in today's organizations, having a direct impact on the following aspects: providing data visualization tools, improving organizational decisions, supporting analysis of breaking organizational information barriers, influencing strategic business decisions, and helping to give meaning to organizational data [6].

When designed to measure performance and backed by a business-oriented BI infrastructure, dashboards enable health managers to be able to measure performance, monitor KPI by preventing deviations, understand undesirable behaviors and redefine the trajectory of the objectives [7]. The dashboard is a mechanism that effectively and efficiently aggregates relevant information to be explored through three tasks: Monitor, Analyze and Manage [8].

Currently, the growing need for hospital organization to have valid and reliable tools that help in the process of analyzing and evaluating the services provided to patients [9] plus the need to have adequate and available information to facilitate measurement of the productivity of hospital organization [10] call for further investigation in this domain. Moreover, international organizations such as the Organization for Economic Co-operation and Development (OECD) launched in 2000 the first document to help simplify the process of productivity analysis [11] which reinforce the relevance and complexity of the domain.

This research is focused on the development of a dashboard for hospital organizations productivity which intends to help to solve the problems identified at Table 1.

Table 1. The key problems of the study

ID	Description
P1	The two problems the authors are trying to help to solve, are: healthcare needs to find ways to explore, analyze, and synthesize valuable information to make decisions in real-time. Sharpe, A., Bradley, C., & Messinger, H. (2007)
P2	and to obtain a productivity analysis that allows to understand that it can be improved in the provision of service to the patients and consequently to obtain a better knowledge of the costs of a hospital organization. Al-HAJJ, S., Pike, I., & Fisher, B. (2012)

This study is based on a comparative analysis between two decision support solutions an institution of the national health service, aiming at the development of a dashboard with a focus on productivity. The measurement of production always had enormous complexity because healthcare is not tradable. This fact hinders the observation of prices and results, technological advances also increases complexity of price analysis and results [11].

The remaining document is organized as follows: chapter two presents the Theoretical background; chapter three, explains the methodology followed in this research (DSR); chapter four presents the first DSR iteration; Chapter five presents the second DSR iteration; finally, in Chapter six, conclusion and future work are drawn.

2   Theoretical Background Review

The article is based on four important areas that will be the subject of a brief analysis in this section Production and Productivity, Key Performance Indicators (KPIs), Dashboards and Visualization and Drill Techniques.

2.1   Production and Productivity

Productivity growth in health organizations is generally lower than the growth of the economy as a whole, which is attributed by experts to measurement problems [12].

Productivity is a critical aspect for the performance of health systems, which can be defined by physical inputs used (labor, capital and supplies) to achieve a certain level of health outcomes in the treatment of a specific disease [13]. Measuring production and productivity is essential to achieve a more efficient allocation of resources in a hospital organization [11]. By measuring productivity, it will be possible to make improvements in the service of organizations, which will not imply an increase in expenses, but an optimization of re-courses [13].

Health organizations need to improve the mechanisms for measuring and analyzing estimates so as to measure productivity, which will improve the performances of the same organizations [11].

2.2   KPI

Key Performance Indicators (KPIs) are undoubtedly critical for the transformation of raw data (numbers) into information or decisions (indicators) [14]. In that sense, several health organizations develop KPI to be able to monitor, measure and manage the performance of their health system and thus ensure efficiency, equity and quality of services provided to their users [15].

Dashboards, when designed to measure performance and backed by a business-oriented BI infrastructure, will enable health managers to be able to measure performance, monitor KPI by preventing deviations, understand undesirable behaviors and redefine the trajectory of the objectives set [7].

2.3   Dashboards

The role of the Dashboard has been increasing in both the health sector and the research community that addresses this topic. Dashboards to be effective in decision support need to strike a balance between the visual aspects and the information contained in them. This equilibrium is intended to avoid the excess of information that is not relevant, providing access only to the crucial information for decision-makers [16].

Dashboards allow reducing the time spent in a manual analysis and facilitate the obtaining of information by a greater number of people, due to the appealing way of presenting the data [17].

Moreover, the idealization of the visual design of a dashboard is a determinant factor for the success or failure of this. The decision makers need of dashboards to help in the decision support phase in organizations. The great challenge is to make use and visualization to facilitate the extraction of the information contained in the dashboard [8].

The dashboard that are developed based on techniques should provide unambiguous information, the visualizations should not be prone to misinterpretations, the information should be ready to consume by the decision makers only so dashboards will be an added value [18]. Such visualization techniques may allow the stakeholders to obtain basic knowledge [19].

2.4 Visualization and Drill Down Techniques

Visualization. Several advantages can be noted with the application of visualization techniques. For example, they help in the exploration of information and provide the tools needed to create a dashboard that provides a big picture [19]. Plus, dashboards based on visualization techniques allow stakeholders to answer their own questions about indicators in a given area and trigger new research that helps to increase the knowledge base, improve existing indicators and also contribute to the emergence of new indicators [20].

Visualization techniques may also enable the creation of visions tailored to the needs of each stakeholder, help increase the analytical, temporal and geographic capacity of the information available to make the best decisions and thus contribute to a more comprehensive analysis and support the authority of health [19].

The dashboard associated with visualization techniques allows the stakeholders to obtain basic knowledge about the disease situation in various health organizations and spread to different regions [20].

Therefore, health organizations using visualization techniques to create dashboards provide a tool that helps improve the understanding of information about services, patients, increasing awareness and improving decision making and, in turn, in patient care [16].

Drill Down. Detailing is a feature of extreme importance that adds more value to the dashboard this idea is defended by some authors. Using the dashboard technology properly, that "A single page is rarely sufficient to present all the relevant performance metrics and therefore the dashboard must have a drill down capability" [21] (p. 18). Furthermore, Park also refer "The drill down may follow the organizational hierarchy from the health system to a business unit hospital, a service, a department and a division, all the way down to individual practitioners" [22] (p. 295).

The drill down technique, when implemented in dashboards, provides stakeholders with intelligent analysis because of the level of detail they can ensure, they can even produce a granular level of information through various techniques such as filtering and zooming [7]. Additionally the drill down technique enables the capacity to analyze indicators, to present answers to the questions of the decision makers and to support the creation of multiple types of perspectives with more or less detail, which enables the materialization of the big picture of that information [10, 23, 24].

3   Research Methodology

The Research Methodology that endowed in the study and is Design Science Research (DSR), this methodology premise to design, build and evaluate the prototype that we are going to want. When the aim of the research is to expand the limits of human capacities and organizations, in order to create new artefacts invoking the Design Science Research Methodology (DSRM) is the right choice [25]. In contrast to other paradigms of research, this one stands out as it tries to develop and obtain artefacts that are possible to buy the effectiveness of this in the real world [26]. The approach (DSR) would include three elements: conceptual principles that help define DSR, practical rules for DRS impersonation, and a prosthetic to perform and conduct research [26]. The DSR is methodology that is known to be an interactive process as it is possible to verify in Fig. 1.

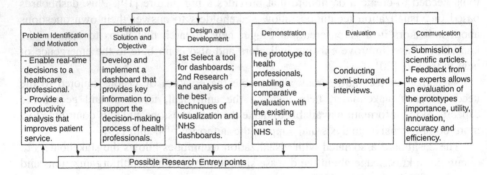

Fig. 1. DSRM process model followed

3.1   Principles

The principles of DSR have their pillars in the engineering of artificial things and information systems are a perfect example of artificial systems, when implementing an objective information system is to reduce the efficiency of the organization [27]. For this reason, the authors follow the principles present at Table 2.

The principles alone do not provide for the requirement of requirements to obtain an effective job in Design science, so here we present the guidelines that have this purpose to establish guidelines that help the research community understand this need [25]. Table 3 shows the 7 DSR guidelines.

Table 2. Design science research principles

Abstraction	This research presents a proposal for a dashboard that intends to improve the decision support process for stakeholders. For this the authors follow indicators that are in the National Health Service (NHS), which is the entity by which the Portuguese Government regulates health in Portugal. The dashboard was validated in a large Portuguese hospital
Originality	The proposed artefact is not in any body of knowledge (BoK)
Justification	The justification for the device is based on the methods proposed for its evaluation
Benefit	The Development of dashboards that allows to obtain in a single location the macro information of production and productivity of a hospital organization, that allows the decision makers to obtain useful information so that they can make the right decisions in a timely manner. This added value can help improve the care of the users and in the last instance may even have an impact on the lives of the users

Table 3. Design science research guidelines [24]

Guideline 1: Design as an Artefact
The artefact proposed by the study is production dashboard and productivity.
Guideline 2: Problem Relevance
Need to have a dashboard on production and productivity that allows an analysis of each hospital organization
Guideline 3: Design Evaluation
Semi-structured interviews
Guideline 4: Research Contributions
A new artefact not present in the body of knowledge. The main principles, practices and procedures of DSR were adopted, in order to increase the credibility of the artefact and the consequent contribution of the study
Guideline 5: Research Rigor

Construction:
    Stephen Few: guidelines practical Rules for Using Colour in Charts
    Gestalt theory: and forming principles of visual perception
Evaluation: interviews with health stakeholders

Guideline 6: Design as a Search Process
The result obtained is the departure from unknown. Combination of good visualization practices and other relevant guidelines for prototype development
Guideline 7: Communication of Research
Evaluated by health stakeholders who have to make decisions using dashboards, which assigns greater credibility. Plus, the submission of the article to a journal/conference with great credibility and respect in the scientific community

4   DSR First Iteration

The authors followed the DSR methodology and carried out 2 iterations to improve the developed prototype. This was populated with information from the hospital in which the interviews were conducted. In this section the authors provide the main information regarding the three phases of each iteration: the proposal; demonstration and evaluation.

4.1   Proposal

The develop the artefact the authors carried out three main steps: research and analysis, ETL processes and construction of dashboards.

The research and analysis phase were divided into two parts. The first part was devoted to researching and analyzing the existing supply of dashboards for health organizations, then a survey on visualization techniques was done to validate what best practices in this area and also to verify if these techniques are already being used in health organizations. The second part of the research and analysis phase was devoted to the analysis of NHS panels and to the study of information and data contained in monthly monitoring and benchmarking. After this study, the authors selected the indicators of the production and productivity segment.

In the ETL (Extract Transform Load) process phase were extracted the data of production and productivity of the site NHK. An analysis of the data and some necessary adjustments were made to import the data into the application.

During the development phase of the panel was made the creation of new measures and indicators that were not created during data exportation. Therefore, new graphic decisions were made having into consideration the previously presented visualization techniques and the guidelines presented in the next section. Figure 2 helps to demonstrate what has been described.

4.2   Guidelines

The use of colours in a dashboard can be a plus, but for that, you need to know the issues related to colour perception and follow rules that help you convey the information. To achieve a dashboard according to good colour practice, there are rules that must be used and followed in the proposed panel, these rules are presented and explained in books [28, 29].

Gestalt's principles of visual perception help to understand which elements of training are crucial to the transmission of information and which elements are pollutants and/or enhancements. These principles are approached by several authors can be studies of more detailed form in one of the following works [28–30].

For the first iteration, a panel was created in the production and productivity segment of the following areas of surgery, internally, external consultations and emergencies. The segments present in the dashboard are defined in a decree-law by the

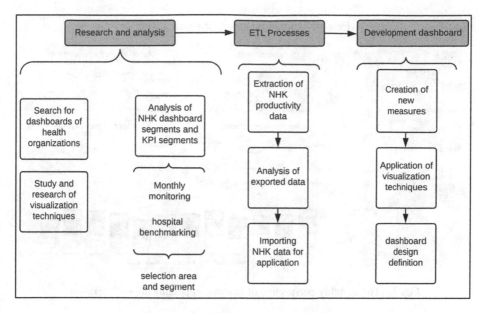

Fig. 2. Proposal development workflow

Portuguese State and are used by the NHS in the benchmarking and the monthly monitoring that this entity provides to hospital organizations. The data contained in the dashboards are referring to the hospital organization where we were to carry out interviews which prematurely stakeholders a familiarization with the data and get faster to evaluate the value of the proposed dashboards. At the level of the dashboard structure this is divided by three types of principal and applied views whenever the data foresee, annual, monthly and target. These perspectives are presented whenever the data allow. In the monthly perspective Fig. 3 it is possible to verify that if it is divided into two header and detail areas being that in the header is where the title of the segment that is being viewed return button and filtering per year is. The detail area is divided in two, and on the left side we have the KPI's with annual values and compared with the homologous year, and on the right side we have the graphs with the monthly distribution of KPIs and other analyzes that can be used by the decision-makers, also in these graphs is the comparison of current year with the homologous.

In Fig. 3, it is noticeable that the authors followed the following principles of Gestalt: Proximity, Similarity, Closing and Connection. These forms are also followed in the views (Annual and Target). Throughout the dashboard the guidelines of the colours rule, being that. The rules 4 and 5 are the ones that stand out more whereas the colours only change when they are different subjects and the use of more attractive colours for the KPI's. In Fig. 4 is referring to the annual information, the color guidelines are also visible.

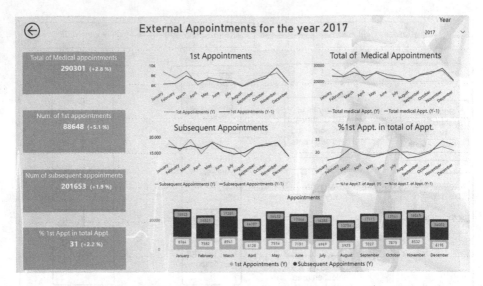

Fig. 3. The monthly perspective of external appointments - 1ˢᵗ iteration

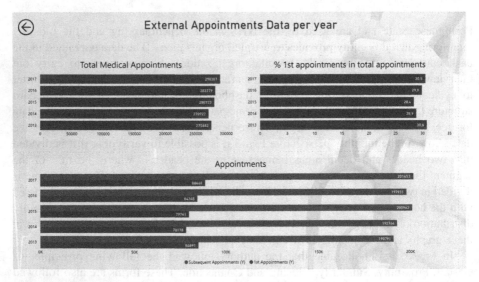

Fig. 4. The annual perspective of external appointments - 1ˢᵗ iteration (Color figure online)

Finally, the separated from the target Fig. 5 only those indicators that have a defined goal in the contract program (annual agreements between the Portuguese State and the hospitals where the levels of economic and financial assistance that are assured by each hospital are defined).

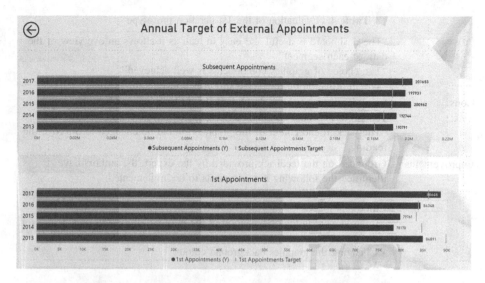

Fig. 5. The annual target of external appointments - 1st iteration

The dashboards are idealized in the light of the guidelines we mentioned before at the level of organization and distribution of information, as well as the choice of graphics and colour choices. Its main objective is to provide ready to use information for decision-makers.

4.3 Demonstration

In order to demonstrate that the artefact develop can be applied in practice the authors have populated the artefact with the respective information of the health organizations under study. The presentation of the panel in terms of navigability, available information, graphics and functionalities allows for a more detailed analysis of KPIS. More information about the evaluation of the artefact can be seen in Sect. 4.4.

4.4 Evaluation

The first evaluation was carried out by a stakeholder in the department of planning, studies, analysis and control. The interviewee was a female with more than 23 years of experience in the hospital and a total of 23 years in the health area.

To present the qualitative results of the interviews, the authors organized the information in a table that shows the pros, cons and improvements. The improvements were the starting point for changing the panel to the next interaction (Table 4).

Table 4. Evaluation of the 1st iteration prototype

Pros	"The dashboard is useful and easy to read as it allows an overview of the data of each segment" "The choice of graphics components is well achieved" "The information on the dashboard only concerns the hospital"
Cons	"The data present in the dashboard should be organized from the more general to the more specific." "The background colour of the graphics should be changed to improve the reading."
Improvements	Grounded on the feedback provided by the expert, the authors have identified the following improvements to be implemented: - creation of KPI's graph that allows the comparison of the current year with the homologous - improvement in the organization of the information of KPI plus macro for the more micro - Changing the graphics background colour

5   DSR Second Iteration

After the first iteration and with the feedback provided by the practitioner the authors performed some changes in the dashboard. At this stage also began the creation of Big picture. The big picture allows you to get a global picture of production and productivity and in case you need the drill down on the dashboards for more detail [16].

5.1   Proposal

With the evaluation of the first iteration the order of the indicators was altered, the background colours were modified, a big picture was started and modified and changed to the areas of the KPIs, with a comparison of the current year with the target. The guidelines that the authors indicated are well visible in this iteration, only added the essential and a care with the colours.

By analyzing Fig. 6 it is also visible the implementation of other principles of Gestalt Enclosure in the KPIs column.

Figure 9 presents a sketch of the Big Picture that is being built with stakeholder input.

In Figs. 7 and 8 it is again apparent that the improvements are implemented and that the colors are to be respected in particular the choice of the background color which only makes a little contrast but improves the reading of this.

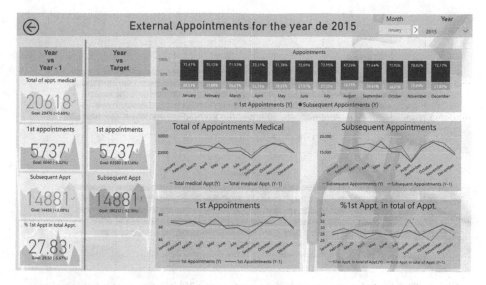

Fig. 6. The monthly perspective of external appointments - 2nd iteration

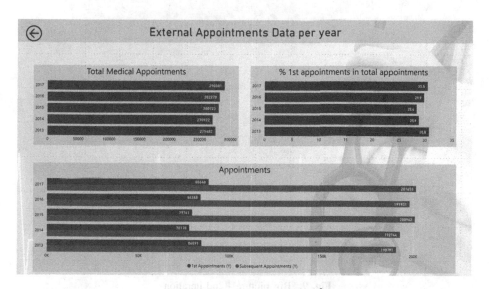

Fig. 7. The annual perspective of external appointments - 2nd iteration (Color figure online)

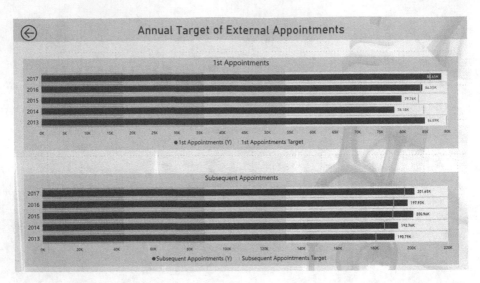

Fig. 8. The Annual target of external appointments - 2nd iteration (Color figure online)

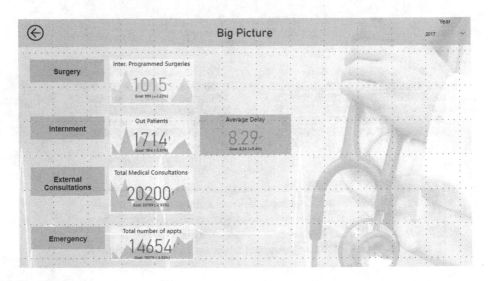

Fig. 9. Big picture - 2nd iteration

5.2   Evaluation

The information used for building the dashboard is originated in the hospital's data sources. This enables an improved decision support.

The second evaluation was performed by a female stakeholder responsible for the surgical area with 41 years old and with a total of 20 years of experience in the area of health both in public hospital organizations and private.

In order to present the results of the interviews, a table was composed of pros, cons and improvements. The enhancements were the starting point for changing the dashboard to the next interaction (Table 5).

Table 5. Evaluation of the 2nd iteration prototype

Pros	"The dashboard allows immediate analysis and monitoring of production data and hospital productivity" "Good colour pallet helps with information readings" "The dashboard displays only hospital information unlike the NHS"
Cons	"It is a macro view of production and productivity"; "It was a plus if the dashboard allowed for a detailed analysis by department, services and stakeholders." "Implement the target of the emergency" "Implement information of the day hospital"
Improvements	Obtain data at the contractual level Allow for a more detailed analysis of the information, allowing an analysis by department, service and health stakeholders Add more emergency information and put information of the day hospital

6 Conclusion and Future Work

The results obtained with the interviews show that the implementation of dashboards on production and productivity is an asset for hospital organizations. Another important aspect that can be concluded from the study is that applying the guidelines of colours and Gestalt's principles of visual allow enhancing the value of dashboards for stakeholders. It is also important to note that the prototype was populated with real hospital information which allowed the stakeholders to become more familiar with the data and to gain more certainty of the information they need to obtain with dashboards. The implementation of the dashboards has all the conditions to materialize since the feedback of the stakeholders interviewed indicate this. Moreover, the dashboard allows us to respond to a need to be able to measure the production and productivity of the hospital organization.

Grounded on such findings and results the authors argue that the two problems identified in Sect. 1 (P1 and P2) were successfully addressed in our proposal.

However, this research also has some limitations. The study was only performed in one hospital which makes it difficult to generalize the conclusions drawn. Plus, only two stakeholders of the hospital were interviewed so far which means that more improvements may very well be proposed in the future.

As a proposal of future work, the authors intend to continue the study with the realization of more iterations in the same hospital and in other hospitals as well. This will premise the generalization of the conclusions. In the second iteration it was requested to increase the detail of the dashboard, so it can provide information by department and service. That's also part of the future work.

Acknowledgements. Fundação para a Ciência e Tecnologia trough the project ISTAR UID/ MULTI/4466/2016.

References

1. Rahimi, H., Kavosi, Z., Shojaei, P., Kharazmi, E.: Key performance indicators in hospital based on balanced scorecard model. J. Heal. Manag. Inf. **4**, 17–24 (2016)
2. Koumpouros, Y.: Balanced scorecard: application in the General Panarcadian Hospital of Tripolis, Greece. Int. J. Heal. Care Qual. Assur. **26**, 286–307 (2013). https://doi.org/10.1108/09526861311319546
3. New Engl. J. Perspect. **363**, 1–3 (2010). https://doi.org/10.1056/nejmp1002530
4. Kawamura, T., Kimura, T., Tsumoto, S.: Estimation of service quality of a hospital information system using a service log. Rev. Socionetwork Strateg. **8**, 53–68 (2014). https://doi.org/10.1007/s12626-014-0044-x
5. Berler, A., Pavlopoulos, S., Koutsouris, D.: Using key performance indicators as knowledge-management tools at a regional health-care authority level. IEEE Trans. Inf. Technol. Biomed. **9**, 184–192 (2005). https://doi.org/10.1109/titb.2005.847196
6. Safwan, E.R., Meredith, R., Burstein, F.: Business Intelligence (BI) system evolution: a case in a healthcare institution. J. Decis. Syst. **25**, 463–475 (2016). https://doi.org/10.1080/12460125.2016.1187384
7. Ghazisaeidi, M., Safdari, R., Torabi, M., Mirzaee, M., Farzi, J., Goodini, A.: Development of performance dashboards in healthcare sector: key practical issues. Acta Inform. Medica. **23**, 317–321 (2015). https://doi.org/10.5455/aim.2015.23.317-321
8. Eckerson, W.W.: Performance Dashboards (2005)
9. Daley, K., Richardson, J., James, I., Chambers, A., Corbett, D.: Clinical dashboard: use in older adult mental health wards. Psychiatrist **37**, 85–88 (2013). https://doi.org/10.1192/pb.bp.111.035899
10. Gordon, J., Richardson, E.: Continuous improvement in the management of hospital wards: the use of operational dashboards. Int. J. Manag. **30**, 414–417 (2013)
11. Sharpe, A., Bradley, C., Messinger, H.: The measurement of output and productivity in the health care sector in Canada : an overview. In: The Measurement of Output and Productivity in the Health Care Sector in Canada: An Overview Abstract Résumé, pp. 1–58 (2007)
12. Hill, H.M.: Measuring productivity in bioanalysis. Bioanalysis **4**, 2317–2319 (2012). https://doi.org/10.4155/bio.12.207
13. Baily, M.N., Garber, A.: Health care productivity. Br. Med. J. **333**, 312–313 (2006). https://doi.org/10.1136/bmj.333.7563.312
14. Staron, M., Meding, W., Niesel, K., Abran, A.: A key performance indicator quality model and its industrial evaluation. In: Proceedings of 2016 Joint Conference in International Workshop Software and Measurement and International Conference on Software Process and Product Measurement, pp. 170–179 (2016). https://doi.org/10.1109/iwsm-mensura.2016.9
15. Khalifa, M., Khalid, P.: Developing strategic health care key performance indicators: a case study on a tertiary care hospital. Procedia Comput. Sci. **63**, 459–466 (2015). https://doi.org/10.1016/j.procs.2015.08.368

16. Zhang, X., Gallagher, K., Goh, S.: BI application: dashboards for healthcare. In: 17th Americas Conference on Information Systems 2011, AMCIS 2011, vol. 5, pp. 3898–3902 (2011)
17. Stadler, J.G., Donlon, K., Siewert, J.D., Franken, T., Lewis, N.E.: Improving the efficiency and ease of healthcare analysis through use of data visualization dashboards. Big Data 4, 129–135 (2016). https://doi.org/10.1089/big.2015.0059
18. Martin, N., Bergs, J., Eerdekens, D., Depaire, B., Verelst, S.: Developing an emergency department crowding dashboard: a design science approach. Int. Emerg. Nurs. 39, 1–9 (2017). https://doi.org/10.1016/j.ienj.2017.08.001
19. McLeod, B., et al.: Matching capacity to demand: a regional dashboard reduces ambulance avoidance and improves accessibility of receiving hospitals. Acad. Emerg. Med. 17, 1383–1389 (2010). https://doi.org/10.1111/j.1553-2712.2010.00928.x
20. Al-Hajj, S., Pike, I., Riecke, B.E., Fisher, B.: Visual analytics for public health: supporting knowledge construction and decision-making. In: Proceedings of Annual Hawaii International Conference on System Science (2013). https://doi.org/10.1109/hicss.2013.599
21. Baskett, L., LeRouge, C., Tremblay, M.C.: Using the dashboard technology properly. Heal. Prog. 89, 16–23 (2008)
22. Park, K.W., Smaltz, D., McFadden, D., Souba, W.: The operating room dashboard. J. Surg. Res. 164, 294–300 (2010). https://doi.org/10.1016/j.jss.2009.09.011
23. Silva, P., et al.: Hospital database workload and fault forecasting. In: 2012 IEEE EMBS Conference on Biomedical Engineering and Sciences, IECBES 2012, pp. 63–68 (2012)
24. Santos, M.Y.: A data-driven analytics approach in the study of Pneumonia's fatalities (2015)
25. Hevner, A.R., March, S.T., Park, J., Ram, S.: Design science in information systems research. MIS Q. 28, 75–105 (2004). https://doi.org/10.2307/25148625
26. Peffers, K., Tuunanen, T., Rothenberger, M.A., Chatterjee, S.: A design science research methodology for information systems research. J. Manag. Inf. Syst. 24, 45–77 (2007). https://doi.org/10.2753/MIS0742-1222240302
27. De Sordi, J.O., Meireles, M., Sanches, C.: Design science aplicada às pesquisas em administração: reflexões a partir do recente histórico de publicações internacionais. RAI - Rev. Adm. e Inovação 8, 10–36 (2013)
28. Few, S.: Information Dashboard Design - The Effective Visual Communication of Data (2006)
29. Wexler, S., Jeffrey, S., Cotgreave, A.: The Big Book of Dashboards: Visualizing Your Data Using Real-World Business Scenarios. Wiley, Hoboken (2017)
30. Knaflic, C.N.: Storytelling With Data: A Data Visualization Guide for Business Professionals. Wiley, Hoboken (2015)

Author Index